SUGAR
Glycerin
SOPO

Fascinating Science Experiments for Young People

GEORGE BARR

Illustrated by
Mildred Waltrip

DOVER PUBLICATIONS, INC.
New York

Copyright

Published in Canada by General Publishing Company, Ltd., 30 Lesmill Road, Don Mills, Toronto, Ontario.
Published in the United Kingdom by Constable and Company, Ltd., 3 The Lanchesters, 162–164 Fulham Palace Road, London W6 9ER.

Bibliographical Note

This Dover edition, first published in 1993, is an unabridged republication of the work first published by McGraw-Hill Book Company, New York, 1961, under the title *More Research Ideas for Young Scientists*.

Library of Congress Cataloging in Publication Data

Barr, George, 1907–
[More research ideas for young scientists]
Fascinating science experiments for young people / George Barr ; illustrated by Mildred Waltrip.
p. cm.
Originally published: More research ideas for young scientists. New York : McGraw-Hill, 1961.
Includes bibliographical references and index.
Summary: Describes experiments in chemistry, astronomy, biology, meteorology, and other branches of science.
ISBN 0-486-27670-8 (pbk.)
1. Science—Experiments—Juvenile literature. 2. Science—Juvenile literature. [1. Science—Experiments. 2. Experiments.] I. Waltrip, Mildred, ill. II. Title.
Q164.B335 1993
507.8—dc20
93-8111
CIP
AC

Manufactured in the United States of America
Dover Publications, Inc., 31 East 2nd Street, Mineola, N.Y. 11501

Contents

Introduction

If you are a young scientist who enjoys doing experiments to discover your own answers to interesting problems, this book should keep you busy throughout the year.

It is the second book in a series which describes exciting research ideas and also guides you in the use of techniques used by professional scientists. It is not a reading book, but rather a doing book. By learning not to take anything for granted, you build good scientific attitudes.

As you proceed you will realize that no experiment is ever completely finished. Instead, the answers that you will find will only whet your appetite for embarking upon new and inviting explorations.

Each chapter of this book will give you experience in a different field. Here are some of the problems which you can pursue in a spirit of adventure:

Can you make a frog hibernate any time *you* want it to?

Can you devise an alarm which will ring when it starts to rain?

How can you produce the most gigantic soap bubbles you ever saw?

How long can you keep fish and plants alive in a tightly sealed aquarium?

Will a loud noise blow out a candle flame?

How can you make your own chemical tester for acids and bases?

What is the best way to clean copper pennies?

Why can't we see stars in the daytime?

How can you make a very inexpensive working model of the solar system?

By using materials found in most homes you can make instruments of a precision which will amaze you. From a small clamp, or even a nut and bolt, you can build a micrometer which will measure thousandths of an inch. From a magnetic compass and some wire, you can construct a delicate electric meter and use it to measure electricity you get from a lemon. And from a tin can, a balloon, and a broken pocket mirror you can design a most intriguing "scope" for studying sound wave patterns.

The research experiments in the following pages are fascinating to do in your home. But they can also be performed and discussed in school, especially for a science club, science fair, or a science assembly show.

Every experiment contains introductory background material. Suggestions are also given for finding out more about the subject. This is as it should be, for a true scientist never stops learning and asking WHY.

Getting Off to a Good Start

You can easily get reliable results in your experimental work if you try to develop certain work habits. It is fun to do the research. But it is still more satisfying to feel that your conclusions are accurate. You can get this feeling of confidence by developing a good technique for attacking a problem.

Do you know *exactly* what you wish to find out? If you cannot pin-point your problem, you will find yourself repeatedly getting off the main track.

Even before you start your experiment, have you found out as much as you can about your problem? It may save you much time, money, and energy. It will also help you observe things in a more meaningful way. Go to library books, speak to experts and authorities such as curators of museums. By the way, it is good to work with a friend who shares your interests.

Do you have everything you need? If not, try to use inexpensive substitutes. It is very useful to have a work place at home with some basic tools and supplies. Work neatly, safely, and orderly. Put things back when you are through with them.

Can you foresee certain difficulties and avoid them?

WEED KILLER
CONTROL
NO
WEED KILLER

Plan your research. Can you think of better ways of performing parts of the experiment than those which are given here?

Report accurately what you see and always be on the alert for unexpected happenings. These may lead you to new problems to investigate later. In this way you will experience the thrills of a research scientist. Louis Pasteur stated, "Chance favors the prepared mind."

Learn to take readable notes in an organized manner. Become familiar with the use of tables and graphs. Do not trust your memory—and always label bottles! Also, write reminders to yourself, and tape these to parts of your experiment.

It is not scientific to base a conclusion upon only one experiment. Gather as much evidence as you can by repeating the experiment many times. If possible, do it in different ways. Think of possible sources of error. Invite your friends' criticism. Remember that an experiment does not always turn out as expected. Trying to discover why an experiment "failed" can often be more interesting than doing proven experiments. This is when your ability as a scientist is really tested.

You have no doubt heard of a CONTROL which is often used in an experiment. It is an important part of your research work which enables you to make very accurate comparisons before and after the experiment.

Wherever possible, you should have another setup just like the one on which you are working. All the conditions for both must be *exactly* the same, except for the one thing that you are doing differently to the experimental one. It is like setting up two experiments with just *one* difference between them.

This technique, used by scientists all over the world, lets you compare one thing at a time. You do not have to guess why any change occurred. Of course, not all experiments call for the use of controls. But keep on the lookout throughout this book for places where you can add controls—even where the specific suggestions may not have been included. They will improve your procedure.

For example, suppose you wish to find out whether a certain weed-killing chemical will kill dandelions in a lawn. Pour the weed killer on some dandelions, and in due time the plants die. Is this a good experiment? Of course not, because somebody may ask the obvious question, "How do you know that these dandelions died because of the weed killer? Maybe it was the time of year when most dandelions die anyway. Perhaps it was the lack of water that killed them."

A scientific control should be set up consisting of a nearby group of dandelions of similar size and health. These will not receive the weed killer. Otherwise, all the other conditions should be *identical*—amount of sunlight, temperature, moisture, type of soil, etc.

Now, if the treated dandelions should die, but those in your controlled group remain healthy, you can be more positive about your conclusion that it was the weed killer alone that killed them.

Fascinating Science Experiments for Young People

CHEMISTRY

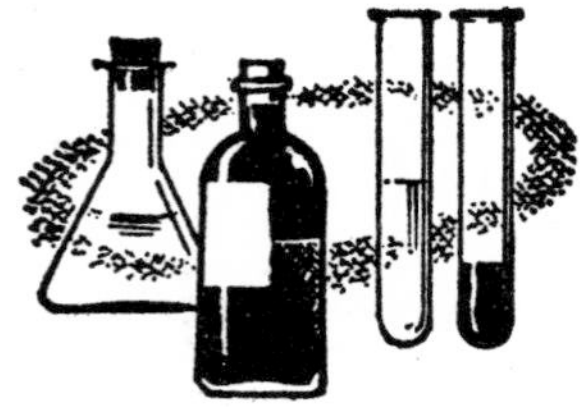

How can you separate colored solutions?

Have you ever watched a motion-picture scene of a chemist at work in his laboratory? Everything certainly looks mysterious and complicated. You get the feeling that it takes many years of special training and practice to know this business.

Sometimes, however, even a highly skilled chemist may use a method which is extremely simple and, at the same time, very effective. Would you like to be a scientific detective and do something that the best chemist could do years ago only after many hours of patient work? By learning a simple technique, you can do the same thing in minutes.

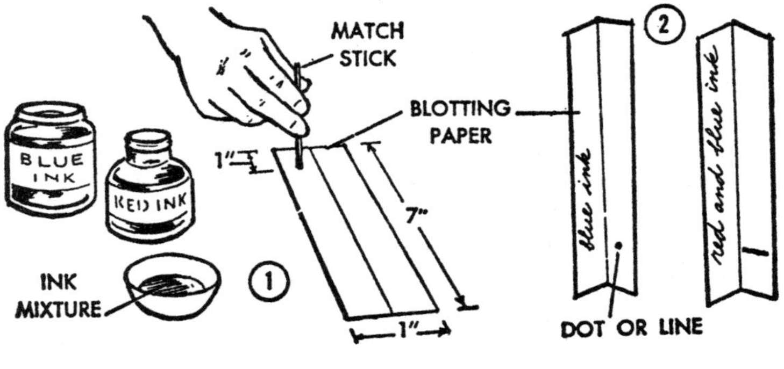

Suppose you want to find out whether a certain colored solution contains just one colored substance or whether there are several colored materials dissolved in it? Does this sound impossible? Well, you can do it. Just make a dot of the solution on a folded strip of white blotting paper, about 1 inch from the end. (See illustration.) After the dot dries, stand the blotting paper in a covered quart-size jar containing about ½ inch of water. The colored dot should be about ½ inch *above* the water.

In a short time the water will creep up the blotter and start dissolving the colored dot. The dissolved colored material in the dot continues to rise until it is redeposited on the paper. You may notice that there is now a separation of colors. For example, a bluish layer may be distinctly visible above a red layer. Perhaps a small mixed area may separate the red and the blue sections. Of course, if the original solution consisted of a single substance in water its color would not change this way.

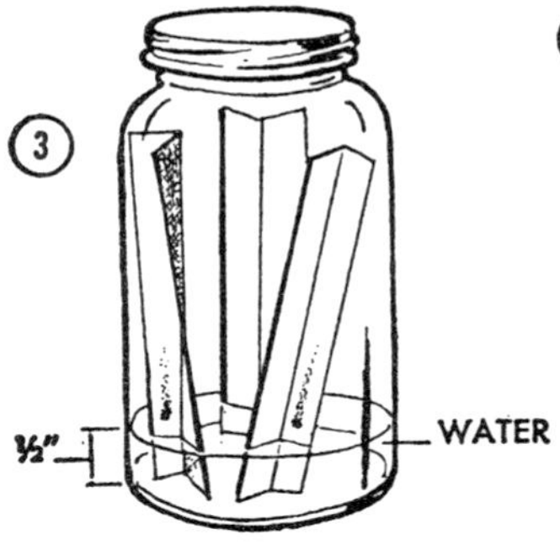

This method of identifying and separating colors which are in solution is known as PAPER CHROMATOGRAPHY (krome-uh-TOG-ruh-fee). The word comes from *chroma* which means "color" and *graphy* meaning "writing." The colored paper strip is called a CHROMATOGRAM. In a way, it represents the fingerprints of the substance.

This technique works on the principle that molecules of different substances travel up the blotting paper at different speeds. The greater the attraction between the blotter and the molecules, the more slowly the molecules rise.

In your experimentation with this fascinating color analyzer you can use only colored substances which are soluble in water, since that is what you have in the jar. Chemists, of course, may use alcohol, benzene, or other dissolving agents. Examples of common water soluble, colored substances you may have at home are: washable ink, red ink, Mercurochrome, laundry bluing, food coloring dyes used for Easter eggs, strong tea, red cabbage, instant coffee, tobacco, ketchup, beet juice, merthiolate antiseptic. No doubt, you will discover more when you start your search.

Make a separate chromatogram for each material you intend to use. These individual strips will act as CONTROLS. By knowing the colors shown by individual substances, then you should be better able to distinguish them when they separate from a mixed solution.

Use white blotting paper, 1-inch wide and about 7-

inches high. Make a heavy pencil line down the center and fold along this line as shown in the illustration. This shape will prevent the paper from bending in the jar. Keep the jar covered, since this creates a humid atmosphere so that the colored solution does not dry on its way up the strip.

To make the dots, use the end of a wooden match or a broken toothpick. Dip this into the solution and then touch the blotting paper. The size of the dot should be a little less than ¼ inch in diameter. However, if you wish, you may make a horizontal line instead of a dot.

For your first analysis, make three separate strips. One should have a dot of only red ink. Another strip should have a dot of *washable* blue ink. The dot on the third strip of blotting paper is obtained from a mixture of several drops of red ink and blue ink prepared in a small dish. Use a pencil to label each strip. Write on the folded side not being used for the test.

For other experiments, mix several substances having distinctive colors. Study the separations by comparing with your controls. Keep good records.

Try using filter paper, or even paper toweling strips, for paper chromatography. You can attach the top of the paper strip to the bottom of the jar cover by means of cellophane tape. If the paper curls, the bottom can be weighted by means of a small paper clip.

Chromatography is often used in crime detection because it gives quick and reliable results. When you learn more about this method of analysis, you will find that it is not limited to colors alone. We used colors for our beginning experiments because it is more convenient and visible. When invisible separations of molecules occur in colorless mixtures, the chemist has ways of developing the strips with chemicals so that different colors appear.

Can you make a tester for acids and bases?

Anyone who has ever worked with a chemistry set has used litmus paper to test for an acid or a base. Litmus paper is inexpensive and can be purchased from any drugstore. It consists of an absorbent paper which contains a natural dye obtained from a lichen, which is a kind of a plant described on page 108. There are two kinds of litmus paper, blue and red.

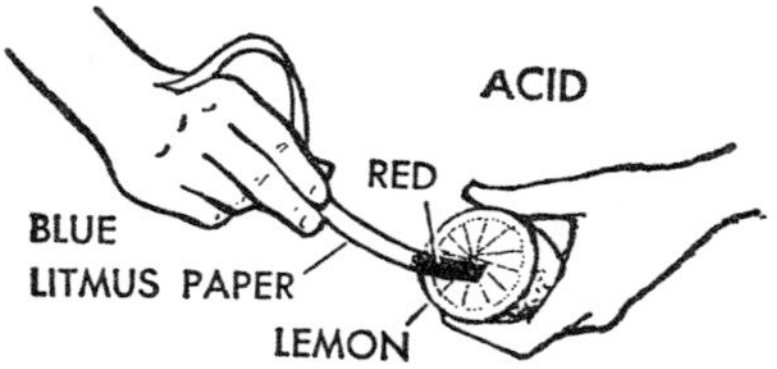

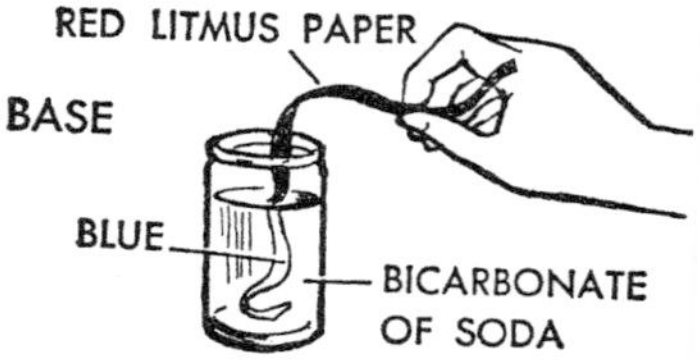

When blue litmus paper is placed in an acid solution it turns red. When red litmus is placed in a base it turns blue. Of course, you know that there are strong acids such as nitric acid and sulfuric acid which can eat away metal, cloth, and shoe leather. However, your home contains many safe acids such as acetic acid in vinegar, citric acid in lemons, and lactic acid in sour milk. The sour taste of many foods is due to acids.

When red litmus is placed in a solution containing a base, it turns blue. A base may be considered a chemical opposite of an acid. In fact, you can neutralize, or weaken an acid by adding a base to it. In the same way, a base can be weakened when an acid is added to it. Bases usually have a bitter taste and feel soapy when rubbed between the fingers. You have many substances at home which are basic, among them are household ammonia, milk of magnesia, and bicarbonate of soda. Bases are also referred to as ALKALIS (AL-kuh-lize).

Touch a moist bit of the food to be tested with the litmus paper. If solutions are used, it is better to employ test tubes. These can be bought in the drugstore. You may also use small glasses, pill vials, or even the glass containers in which toothbrushes are often purchased.

Litmus is called an INDICATOR because it indicates whether something is an acid or a base. It is only one of many natural substances which possess different colors in acids and in bases. You can experiment with the following easily obtainable materials to make your own indicators:

Red cabbage, tea, rhubarb stalks, diluted grape juice, purple blossoms of hollyhocks, red or purple dahlias, purple iris, red petunias, violets, cherries, huckleberries, elderberries. In addition you might try any colored juices from flowers, fruits, or vegetables to discover new indicators of your own.

In general, the procedure is to squeeze, crush, stir, and shake the fibers in some hot water until the water takes on a color. Then see how a small amount of the liquid behaves when either an acid or a base is added to it. The most common household acid for this purpose is vinegar. To test the indicators' reaction to bases, use several drops of household ammonia water or a pinch of baking soda. Finally, bottle your indicators and label them.

For example, here is how red cabbage is converted into an indicator: Chop up a few of the most purple leaves. Boil in a small amount of water for about twenty minutes. Use a small flame. Do not burn by allowing water to boil off. Cool. The liquid is now purple. Bottle some of this and label it: CABBAGE (NEUTRAL).

To a portion of the remaining purple liquid, add just enough baking soda to turn the liquid green. Bottle this and label it: CABBAGE (BASIC). To the last portion of purple liquid, add just enough vinegar to make it turn red. Bottle this and label it: CABBAGE (ACID).

To test whether some unknown liquid is an acid or a base, pour a small amount of the indicator in a test tube. To this, add some of the unknown liquid. Record what happens.

How would each of the three cabbage indicators behave with a substance such as milk of magnesia which is a base? The neutral cabbage will change from purple to green. The acid cabbage will change from red to green. The basic cabbage will not be affected.

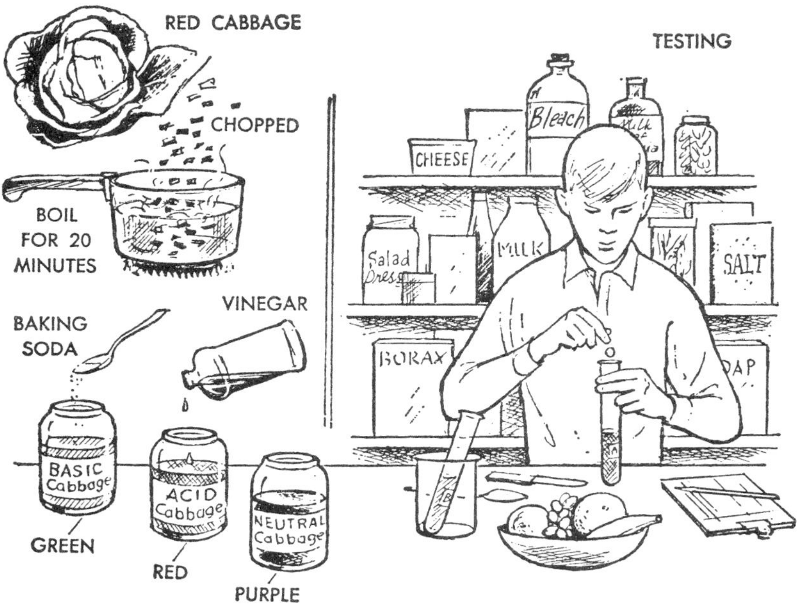

(Actually, the neutral cabbage indicator is unnecessary. It was included here because of the fun in seeing the color change.)

Do you think you can make paper indicators by soaking blotting paper or mimeograph paper in the indicators and then drying them?

Remember that many substances are not acids or bases. They are neutral. Of course, if they are very weak acids or bases and your indicator is not sensitive, then you will also get a neutral reaction.

Test the following substances: orange, lemon, lime, grape, tomato, onion, celery, grapefruit, rhubarb, soda water of any flavor, table salt, chlorine water used for laundering clothes, apple, banana, milk, benzene, vinegar, sour milk, cottage cheese, alcohol, human perspiration, aspirin, epsom salt, starch, washing soda, plaster, pickle, baking powder, vitamin C pill, borax, boric acid, strong laundry soap, lime water, mayonnaise, French dressing.

Use this form for your notes:

SUBSTANCE TESTED	INDICATOR USED	COLOR CHANGES OF INDICATOR	FINDINGS		
			ACID	BASE	NEUTRAL

Can you formulate your own tooth paste?

Many years ago our country was not industrialized and there were no corner drugstores, supermarkets or department stores. Most families had to prepare their own food, design and construct furniture, and weave and sew clothing. In addition, many of the household preparations had to be developed after much experimentation. People proudly talked about their personal formulas for soaps, medicines, and even shoe polish.

You too can have something to be proud of by developing your own inexpensive tooth powder which you and your family will be happy to use. You will also gain much interesting experience as an amateur pharmacist.

Despite the advertisements which appear everywhere,

dentists agree that most tooth pastes and tooth powders serve the same purpose, regardless of price. Their main job is to cause a mild, harmless scouring action which brushes away the particles of food clinging to the teeth.

It is believed that cavities are formed when bacteria use the food particles of especially sugars and starches to produce acids which dissolve away the hard, protective enamel of the teeth.

The following formulas and suggestions are for your experimentation. Mix very small amounts for each trial recipe. Vary the quantities of each ingredient until you have a product which is not too gritty, leaves a pleasant taste, and makes the teeth feel clean. When you are finally satisfied, then you may mix up a larger quantity.

Store each product in a folded piece of wax paper which is properly labeled and secured with a rubber band. Keep careful notes of each experiment.

Your first recipe for a tooth powder is one recommended by dentists as being economical and very effective. It consists of 1 part table salt and 3 parts baking soda, which is also known as bicarbonate of soda. These ingredients are found in every home. Try fine salt in one formula, then use coarse salt in another. You will get accustomed to the salty taste which is easily swished away by several mouthfuls of water.

Another type of tooth powder has a more pleasant taste and enables you to do more experimentation. You will have to purchase a small quantity of powdered chalk from your druggist. This is also known as precipitated

chalk and is very inexpensive. Start your experiment with small amounts of the following formula:

4 parts of precipitated chalk
1 part baking soda
1 part powdered sugar

(Use a very small thimble as a measurer of "1 part.")

In making up different proportions of each ingredient you should know that the chalk is used for its scouring action, while the baking soda is added mainly for counteracting the acids in the mouth. The sugar is included only to provide a pleasant taste, so use it sparingly. Try using a small amount of saccharin as a substitute for the sugar.

You may wish to add a small quantity of powdered white soap. This is what makes commercial tooth powders foam slightly.

If you want a flavor, add a few drops of oil of peppermint, wintergreen, cinnamon, or vanilla. You can get tiny amounts of these from your druggist.

When you find the best combination for the tooth powder you can make a tooth paste from it. Make up a solution of equal parts of glycerine and water. Mix this solution with the powder in a deep bowl until a smooth, thick paste is formed.

You can use milk of magnesia instead of the baking soda as an acid neutralizer, if you wish. You can keep the tooth paste in a jar.

CHEMISTRY—*More to find out:*

How are secret invisible messages written?

"Invisible ink" can be made from certain household liquids such as onion juice, lemon juice, or milk. Use any one of these to write a message on white paper, with a very clean pen point. You can see what you are writing if you place a light on your left side and look at the writing from the right. Dip the pen into the juice often.

Do not scratch the paper. The writing should be invisible when dry.

To read the message, it is necessary to char the writing. Try heating with an electric iron or baking in an oven for a few minutes.

The writing appears because the solids in the juices burn or char faster than the paper does. Experiment with many juices.

What cleans pennies?

Place a tarnished copper penny in a small drinking glass and squeeze some lemon juice over the coin. Does the penny get clean? Pour some table salt into the glass. In a few minutes the penny will be as bright as new. The cleansing action is due to hydrochloric acid which is formed from salt and the citric acid in the lemon juice.

Will salt alone clean pennies? Can other household acids be used with the salt instead of lemon juice? Try vinegar; it contains acetic acid. Can rust be cleaned this way? Can you clean nickels and dimes too?

What kinds of crystals can you grow?

Bring to a slow boil about one-half of a glass of water in a pan. Add a teaspoon or two of alum, obtainable at a drugstore, and stir. When this dissolves, add more. Keep adding alum and stirring until no more dissolves. You now have a *saturated* solution for growing crystals.

Pour the hot solution into a glass, preferably one of Pyrex. Place a pencil across the top and suspend a string

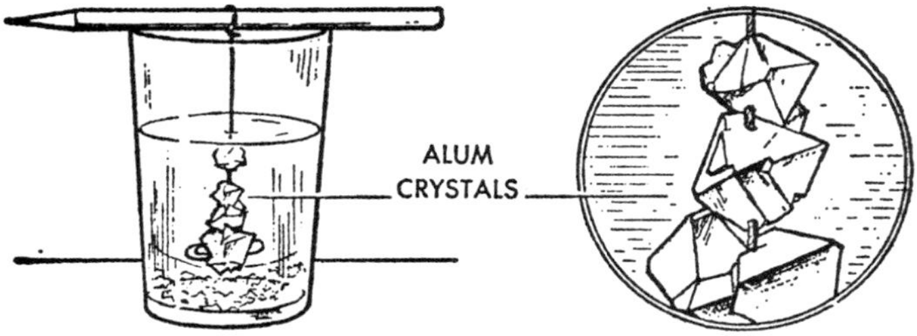

from the middle of it. A paper clip may be tied to the end of the string.

Wrap the glass in a cloth so that cooling takes place *slowly*. This also prevents evaporation. Every day, observe the formation of crystals which are hanging on to the string or resting on the bottom. Do not disturb the crystals and they will grow larger.

Crystals have many different shapes. Grow crystals of table salt, epsom salt, sugar. The crystals can be preserved by coating them with clear nail polish.

What makes silver black?

The black tarnish on silver spoons, coins, and jewelry is formed when silver combines with sulfur. Silver exposed to the air will darken because of a sulfur compound called hydrogen sulfide. It is a gas which is always in the air. It has a rotten egg odor and is produced in industrial plants and when living things decompose.

You can demonstrate that sulfur turns silver black by covering a silver coin with some moist sulfur powder. Most chemistry kits have it. Ask your druggist for a pinch. A rubber band too, when wrapped around a silver coin, will turn it black because sulfur is used in the preparation of rubber. Certain foods such as eggs and mustard contain sulfur. See if they turn silver black.

Where can you see the breakdown of atoms?

You can see atomic energy explosions if you have a magnifying glass and a watch or alarm clock with a luminous dial. Stay in a pitch-dark room for a while to get your eyes adjusted. Then focus the lens sharply on the numerals. You will see continuous light flashes.

You are witnessing an effect of the breakdown of atoms. The luminous paint contains a tiny amount of radioactive substance such as radium. This keeps shooting out particles as it disintegrates. These particles cause a fluorescent material, such as zinc sulfide in the paint, to light up. Experiment with various dials and also different magnifiers, including a microscope.

Can you make a foam-type fire extinguisher?

In some fire extinguishers an acid is added to a solution of bicarbonate of soda. Carbon dioxide is generated which smothers the fire. However, for an oil or gasoline fire, it is desirable to produce a tough blanket of foam which can float on the oil and smother the fire. You can make a thick foam, almost like the commercial extinguishers by doing the following:

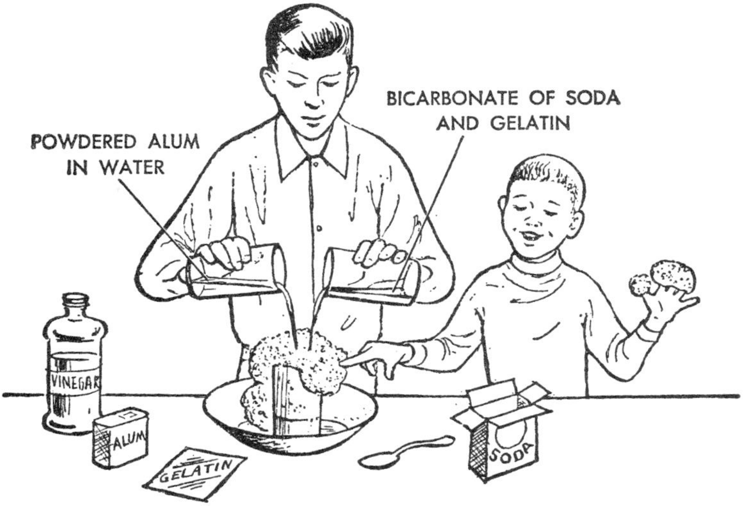

Get two glasses, each one-quarter full of water. In one dissolve a teaspoon of bicarbonate of soda and one-half teaspoon of unsweetened gelatin. In the other glass dissolve a teaspoon of powdered alum.

Pour the contents of both glasses, at the same time, into another empty glass standing in a soup plate. See the large creamy bubbles overflow the glass into the plate. Allow the tough bubbles to cling to your hand.

Vary the proportions and methods of mixing. Are more bubbles produced when some vinegar is added? Compare these bubbles with the non-foaming kind produced when only vinegar is added to the bicarbonate of soda.

The alum acts as an acid, while the gelatin makes the bubbles tougher. In commercial foam extinguishers licorice is used instead of gelatin.

ASTRONOMY

Why can't we see stars in the daytime?

As the earth revolves around the sun during the year, the stars are always up in the sky. Yet we can only see them when the rotation of the earth brings us into the darkened portion of the earth we call night. (See illustration.) During the day the very dim light coming from the stars is lost, in contrast with the strong sunlight or the light reflecting from the particles in the atmosphere.

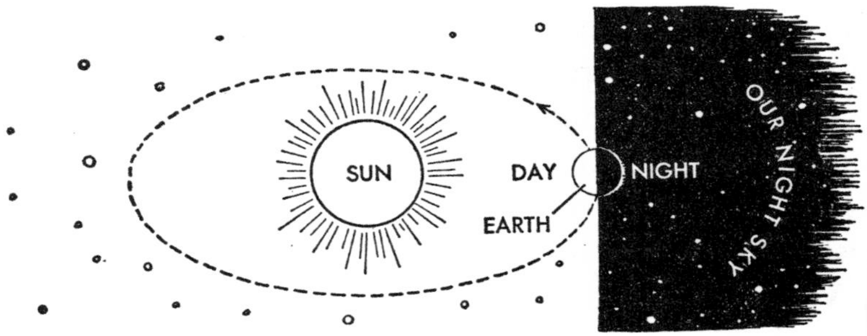

It is a common experience to have dim lights obscured by stronger lights. For example, on a dark road at night a truck, with many small red lights all over the body, may be approaching. When its brilliant headlights shine into our eyes we lose our ability to see anything of the truck except the powerful lights.

Have you ever had someone shine a bright flashlight into your eyes in a dark place? You probably were not able to see the holder of the flashlight. And if the person were smoking you would not be able to see the glowing cigarette or cigar embers.

There is a convincing way to show your friends and classmates that stars cannot be seen when the observer is in sunlight or in a lighted place. Also, that stars can be seen when a bright light no longer shines upon the viewer.

Cut a piece of thin cardboard so that it fits into an ordinary white, mailing envelope. Remove the cardboard and punch it full of holes to represent the stars. You may wish to make it more interesting by making actual constellations, such as the Big Dipper or Orion.

You may use a leather hole-punch or the device for making holes in looseleaf pages. If a nail or large pin is used, make sure that the holes do not have burrs or lifted edges. If they do, press the edges flat.

Replace the perforated cardboard into the envelope. Hold it up toward the window having most light. If you can still see the "stars" through the closed envelope, you must insert white, unruled, writing paper. If one sheet is not sufficient to blot out the holes, then insert another sheet. However, do not use too many sheets—only enough to prevent the holes from showing through the envelope when viewed toward the light. This envelope, with its contents thus prepared represents the "daytime sky." In other words, the stars are there but we cannot see them.

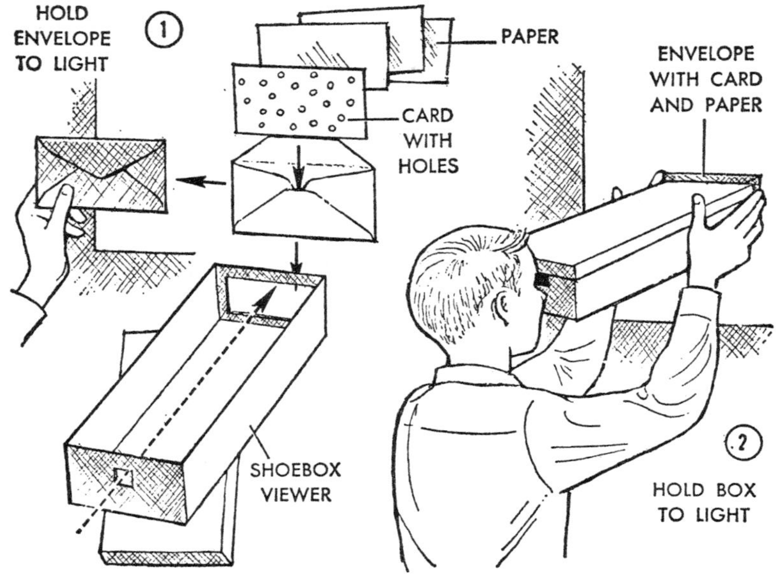

In order to see the "stars" you must darken the side of the envelope facing you. This can be done by looking at the envelope through a mailing tube held close to it. Perhaps you can think of other means.

A very good method is to obtain a shoe box and cut a 1-inch peep hole in one end and a 2-inch by 4-inch hole in the opposite end.

Place the envelope tightly against the larger hole of the box. Look through the peep hole toward the light, and lo and behold!—the stars now become visible!

In explaining this to your friends, emphasize the fact that the light shining through the envelope and the perforated cardboard is not changed throughout the demonstration. In the same way, the light from the stars is always the same, at night, as well as by day. Only the light around the viewer changes.

Perhaps you can devise a method whereby an entire class can see this demonstration at the same time. A larger cardboard might be placed over a window. In the center is an opening where you can place a very large envelope prepared as explained above.

Now, if the room has dark shades, the stars will gradually appear through the envelope as the shades are drawn and "darkness falls."

Can stars help you keep time?

In ancient times, and even today in some parts of the world, intervals of time during the night were indicated by the changing positions of certain stars. You can learn to do this too, and perhaps with greater accuracy. But first, some basic facts in astronomy must be reviewed.

The earth rotates on its axis from west to east, making one complete turn in twenty-four hours. That is why the sun, moon, and the stars seem to move through our sky in the opposite direction—that is, from east to west. The North Star however, remains in about the same position all the time. This is due to the fact that the earth's axis is always pointing toward the North Star. All the other stars appear to rotate around the North Star in a *counterclockwise* motion. This means opposite to the direction the hands on a clock move.

In other words, these stars make a complete circle of

360 degrees once every twenty-four hours. Therefore an imaginary reference line connecting any star to the North Star will move through an arc of 15 degrees in one hour. (360 degrees divided by 24 hours.)

Here is how you can find the North Star and become familiar with it. Face north. If necessary, use a magnetic compass. Point your extended arm toward the north halfway up between the horizon and overhead. There are so few stars in this part of the sky that most likely you will be pointing to the North Star.

It is not a very bright star. It is easy to identify because it is the end star of the handle of the Little Dipper. The two end stars of the bowl of the Big Dipper are called "pointers" and will point to the North Star, also known as the Pole Star and Polaris (poh-LAR-iss).

The pointers represent the hands of a clock, moving counterclockwise 15 degrees in one hour. You now need a device for accurately determining the angle between two separate observations. Obtain a protractor. Attach a 4-inch string with a small weight at one end. When suspended freely, this becomes a plumb or vertical line from which your angles are read.

Your ingenuity will no doubt help you design devices other than the protractor for this purpose. You might try a circular card containing more angles ruled on it. Perhaps with experience, you might invent a quick-reading sliding scale, or something to tighten against your reference point.

To use this instrument, hold it at arm's length toward the North Star. Do it near a street light so you can read

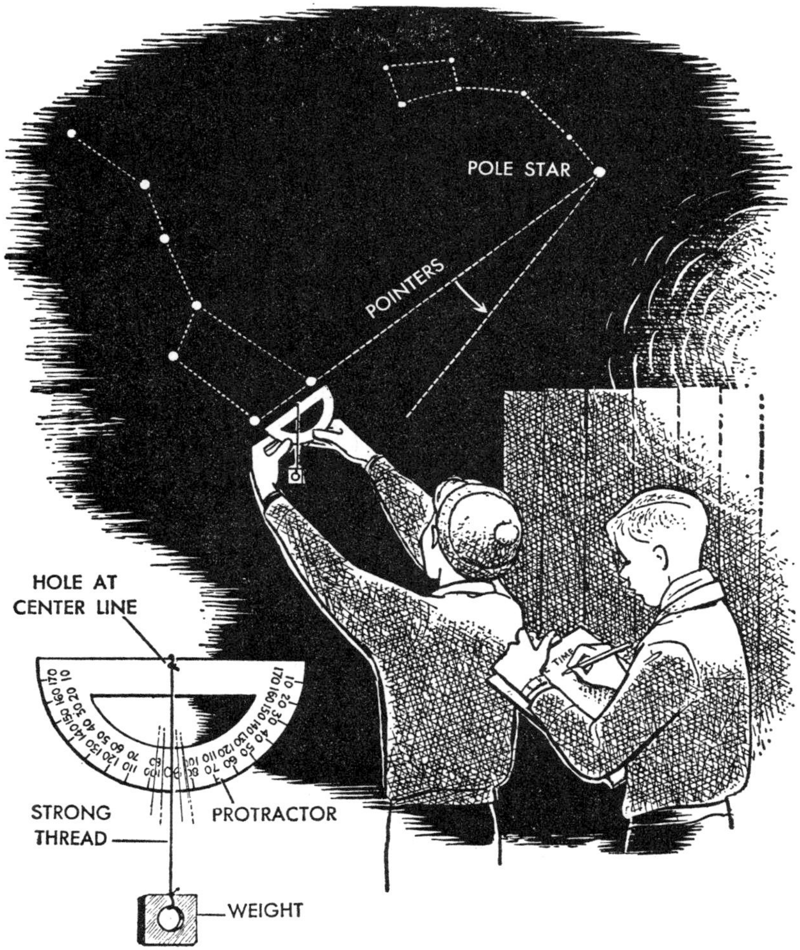

the scale. Line up the top of the protractor with the "pointers" of the Big Dipper. Then, using your thumb and forefinger, grasp the plumb line against the scale. Now read the number the plumb line is covering and record it. Also note the time.

When a similar sighting is repeated at a later time in

the same evening, again read the number under the plumb line and write it down. The difference between the two readings is the number of degrees the "pointers" moved.

This time you can easily calculate the new time without your watch since you know that 15 degrees represents one hour difference. You can also get the time in minutes by allowing four minutes for each degree difference, (obtained by dividing 60 minutes by 15 degrees). This should enable you to get the new time quite accurately, depending upon your technique.

If the Big Dipper pointers are too low for observing, you can use any star close to the North Star which you can recognize again. Make an imaginary line to the North Star. Sight on this line.

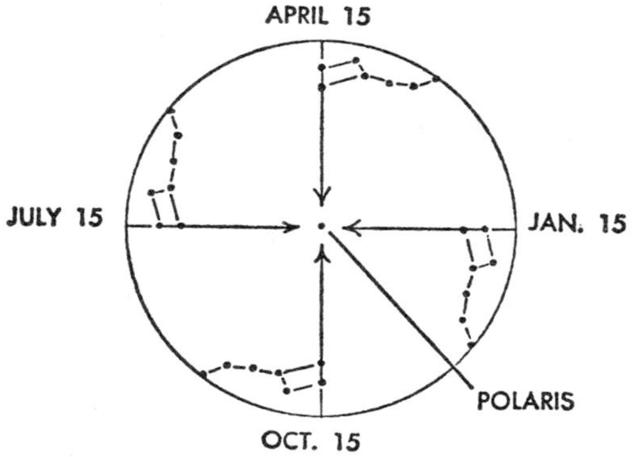

POSITION OF BIG DIPPER AT 9 P.M. THROUGHOUT THE YEAR

Every night the stars come up about four minutes or 1 degree earlier. You can use this fact to carry one reference reading for many days.

Knowing the stars and their positions can give you a

calendar as well as a heavenly clock. The accompanying diagram indicates the position of the "pointers" in the Big Dipper throughout the year. Can you design a card with a plumb line which can be used as a direct reading clock throughout the year?

Where is the sun today?

The sun appears to move across the sky from east to west because of the rotation of the earth upon its axis from west to east. However, if you have been observant, you know that the sun rises exactly in the east and sets exactly in the west on only two days of the year.

One of these is the first day of spring, which is on or about March 21. The other is the first day of autumn, about September 23. These dates represent the spring and autumn EQUINOXES (EE-kwin-ox-ez) when day and night are of equal duration.

After March 21, the sun rises and sets every day a little more to the north until June 21. This is the first day of summer in the Northern Hemisphere. The sun

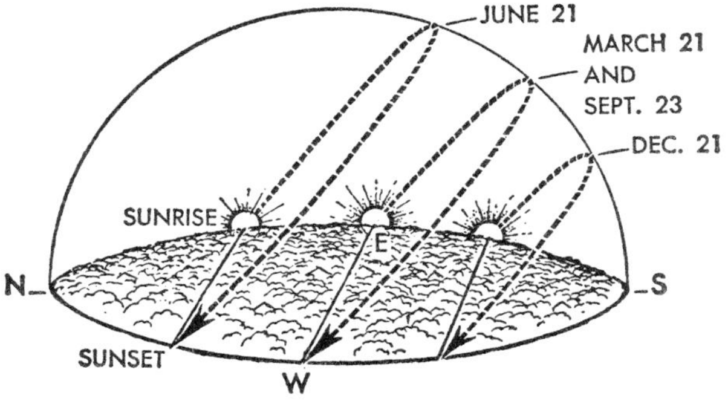

now ceases its apparent northern journey through the sky. It starts coming up and going down every day more to the south until about September 23. Now it again rises and sets exactly east and west.

After September 23, the sun rises every day a little more to the south of east and sets the same distance to the south of west. On December 21, the southerly motion ceases and after this date the sun again rises and sets more to the north every day.

This northerly and southerly shifting of the locations of sunrise and sunset is caused by the yearly revolution of the earth around the sun. The axis of the earth is tilted 23½ degrees from the vertical as it speeds through space. The earth's axis is also parallel to itself all the time and pointing in the same direction toward the North Star.

Because of these conditions the North Pole of the earth is tilted *toward* the sun for six months and *away* from the sun for six months. You can now understand why the sun seems to come up from different points on the horizon during the year.

You can study these interesting changes in the sun's position during sunrise and sunset throughout the year.

A good method is to select an upper window with a clear view of the eastern or western horizon. An ideal view, of course, is over open water. Stand where you can sight along a vertical section of the window. The window lock may also be used for this purpose. Line up this point with another marker in the distance. This can be a pole, tree, chimney, television aerial, side of a

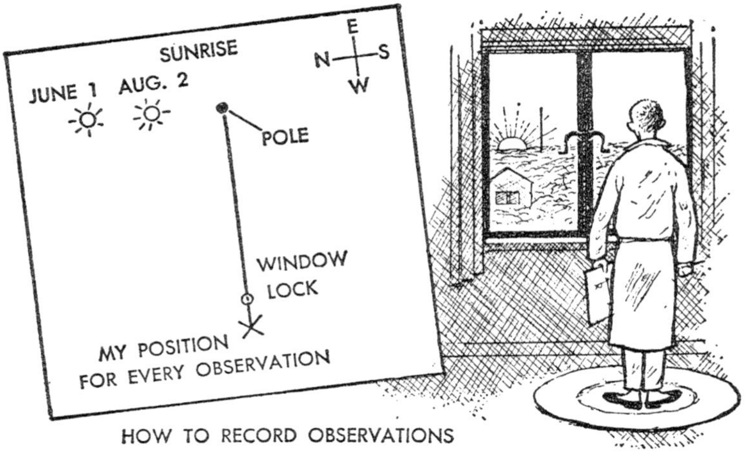

HOW TO RECORD OBSERVATIONS

house, or any landmark you select. Keep records of the date and indicate the location of the sun. You may wish to use the method suggested in the illustration.

You can notice significant changes if you take your sightings at about two week intervals. It is best to do this at sunrise, sunset, or as close to this time as possible. Since the angle changes rapidly as the sun rises, always take your readings when the sun is at the same height very near the horizon. Try especially to take readings on March 21, June 21, September 23, and December 21.

Another method of studying the different locations of the sun in the sky throughout the year is to trace the direction of the shadow cast by a vertical stick at sunrise or sunset.

Another interesting research idea you might wish to try in connection with the sun's apparent movements is to study the length of shadows during the year.

When the sun rises north of east, it makes an arc in the sky which is higher than when the sun rises due east or south of east. When the sun is high the shadows are smaller. Therefore, by comparing the lengths of shadows cast by the same object in different seasons, but at the same time of day, you can learn the relative height of the sun.

Set up a vertical stick about a foot long. Or you may use a flagpole or any vertical object. Measure the length of the shadow exactly at noon. If there is daylight-saving time in your section of the country take your measurements one hour later. If you are confused about daylight-saving time measure the shortest shadow, since this occurs exactly at noon.

Complete the chart and bar graph with your results. Use a period of several months or an entire year.

ASTRONOMY—*More to find out:*

Can you make a working model of the solar system?

First experiment with the following model, using only the sun, earth, and moon. With this as a basis you can invent improved ways of putting the other planets in orbit.

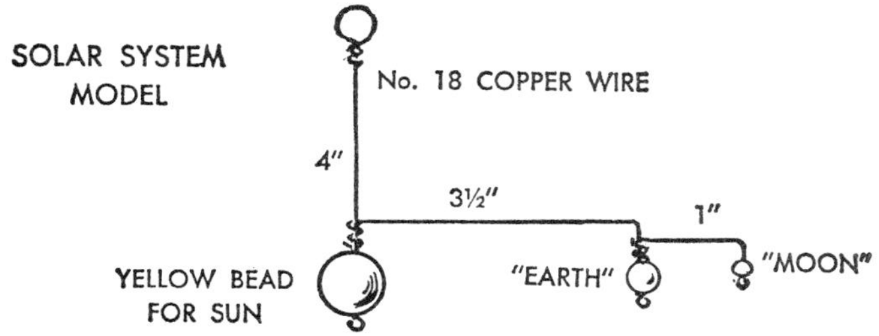

Get about 24 inches of No. 18 bare copper wire, or use bell wire with the insulation removed. This wire can be bent by hand and cut with an old pair of scissors. You will also need a 1-inch wooden bead with a yellow color for the sun. A ½-inch green bead will represent the earth. For the moon, use a small pearl bead, or a rolled up and pierced ball of aluminum foil, about one-fourth the diameter of the earth.

Study the diagram. The small loop on the bottom of each bead prevents the bead from falling through. The end of the wire holding the earth is given at least three turns around the vertical wire holding the sun. This should keep it turning freely at right angles without sagging.

The wire for the moon is given the same treatment. If you find it difficult to twist the extreme ends of the wire, start twisting about 1 inch from the end. This gives you greater leverage. Cut off the excess wire.

To use this model, hold the large loop above the sun in one hand. With the other hand make the earth revolve around the sun. You can also rotate the earth on its axis. In the same way, the moon can be made to revolve and rotate.

Can a watch be used as a compass?

This is a very old method used for finding north if no compass is available. Boy Scouts swear by it, but astronomers and explorers who demand greater accuracy do not think much of its reliability. Find out for yourself how closely this method helps you find true north.

Hold a watch or clock horizontally, with the face up. Stand a thin stick vertically in the center of the watch. Now turn the watch until the shadow of the stick falls along the hour hand. North will be halfway between the shadow and the twelve.

Do this several times a day. Does this method always indicate north in the same direction? How close is it to true north? The best way is to find the North Star at night. Read directions on page 34.

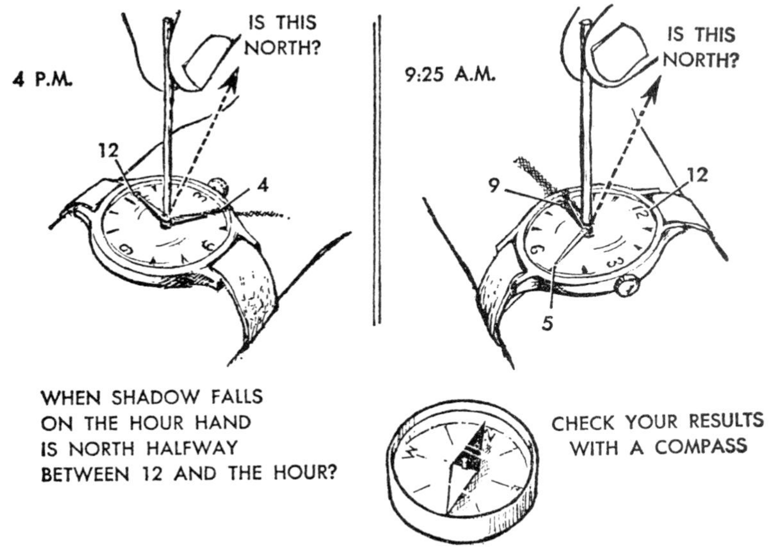
4 P.M.
IS THIS NORTH?
12
4
9:25 A.M.
IS THIS NORTH?
9
12
5
WHEN SHADOW FALLS ON THE HOUR HAND IS NORTH HALFWAY BETWEEN 12 AND THE HOUR?
CHECK YOUR RESULTS WITH A COMPASS

How does north found by the watch method, compare with north obtained with a magnetic compass? Remember that a compass needle points to the North Magnetic Pole, and not to true north which is the Geographic North Pole. These are two different places. Find out how much of a correction has to be made in your locality. Ask your librarian for such a correction chart.

Is the watch compass more accurate during certain seasons? Keep careful records for this research.

How can the moon be photographed?

You do not need an expensive camera to build up a handsome collection of clear photographs of the full moon as well as of other phases.

Pictures must be taken from a place as dark as possible, away from street lights. Set the camera distance to infinity (for objects farthest away). Excellent exposures of a full moon have been obtained at one-fiftieth of a second at f/8 using a medium speed film such as Panatomic X. The exposure would, of course, have to be increased for a crescent moon.

If yours is a box camera just snap the moon at the slowest "instantaneous" speed and the largest aperture. Experiment with several exposures and different films.

An interesting variation is to take a series of individual photographs of the moon upon the same negative every ten minutes. Set the camera upon a tripod or something solid. Do not wind the film after each exposure. You will be rewarded with a picture showing the moon at different positions along its path.

Can you take a good picture of star trails?

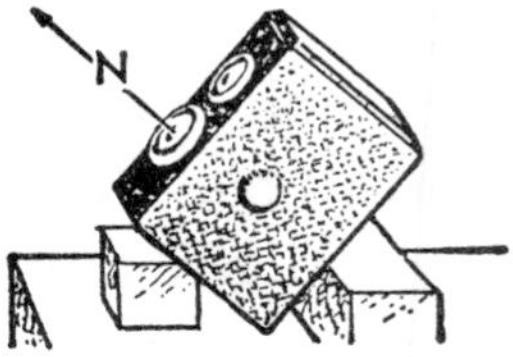

As you know, the North Star remains in about the same place in the sky. Each star near it moves in a circle around the North Star once every twenty-four hours. A time exposure of several hours will show star trails, or arcs of light whose size depends upon the time of exposure. For example, a four-hour exposure will produce an arc which is one-sixth of a complete circle.

For best results, choose a moonless night free from clouds. Keep the camera away from house or street lights. Recurrent airport beacon lights can also ruin your picture. Set the shutter on *time,* distance at infinity, and open the lens as far as possible. Do not walk or run heavily near the camera. Remember to close the shutter when you are through.

It will be interesting to discover that even the North Star moves. The axis of the earth does not point *exactly* to that spot.

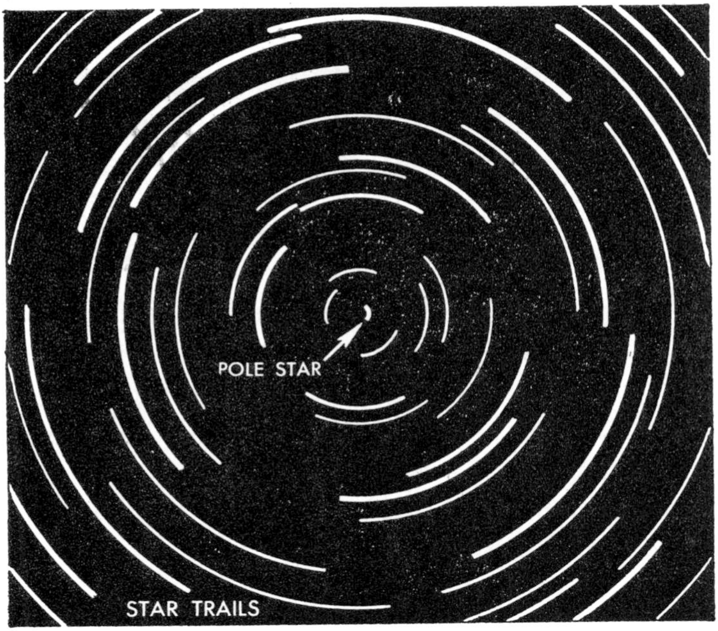

What planets are visible tonight?

You can find out which planets are in the sky at this time of the year by referring in the library to *The Science Newsletter* (the last issue of the month). *Natural History,* a magazine published by the Museum of Natural History in New York City, also contains this information. Certain newspapers and magazines carry monthly star maps and planet guides. Almanacs also have charts. Consult a planetarium near you. Your librarian certainly knows how to find this information.

Is the full moon really larger?

When the full moon first comes up it looks much bigger than when it is overhead. By a simple test, however, you can prove that it is the same size.

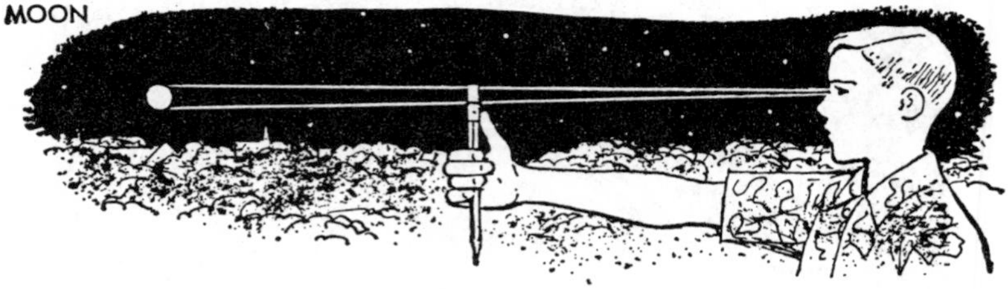

Hold a pencil with a new, large eraser on it in a vertical position at arm's length. See that the eraser covers the moon when it is rising. If the eraser is too large rub some of it off on the sidewalk. When the moon is overhead, compare its size against the eraser.

Can you think of methods to make more accurate observations?

MAGNETISM
AND
ELECTRICITY

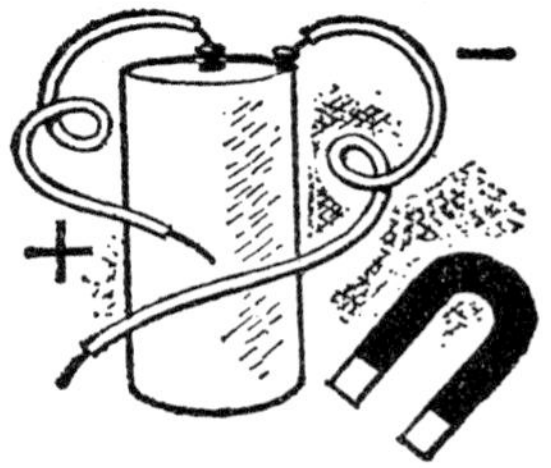

Can you detect small electric currents?

In most of your experimental work with dry cells it is easy to know when current is passing through a wire. You simply make a bell ring or a light go on. But suppose you were dealing with electricity not strong enough to ring a bell or light a bulb?

In scientific laboratories there are ingenious electrical meters which can measure currents many millions of times weaker than those obtained from dry cells. Of course these instruments are exceedingly expensive and often complicated. But with a minimum of expense and materials you can build a very sensitive meter. It will enable you to perform many experiments involving extremely small currents.

This homemade meter is a magnetic compass with thin, insulated copper wire wound around it in the form of a coil. (See illus., p. 48.) When the ends of the coil are connected to a source of current, the compass needle moves from its regular north and south position.

It works because a magnetic field is produced around the wire when electricity passes through the wire. This magnetism makes the compass needle swing. The magnetic effect is made stronger when more current goes through the wire. The instrument is more sensitive to weak currents when there are more (but not too many) loops of wire around the compass, or when the loops are closer to the compass needle.

Obtain a small magnetic compass whose needle swings easily, and can be trusted not to get stuck if the case is tilted slightly. Wind about fifty turns of thin *insulated* wire around the compass in the form of a narrow coil. Tape or tie parts of the coil so that the wires will not get loose. Leave about 6 inches of wire at each end of the coil for making connections later.

The best wire to use is No. 22 cotton-covered or enameled magnet wire. This special thin wire enables you to get all the turns *close to the compass needle.* You can get this wire from old broken bells or radio loudspeakers. (Remember to scrape the ends of the brown enameled wire before making connections.) If nothing else is available, use bell wire.

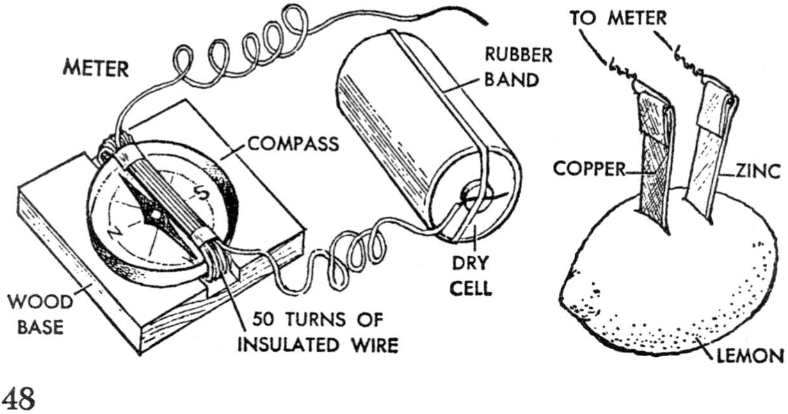

Since the coiled wire makes the compass wobbly, you must devise some way to keep it level. You might attach it to a wooden base in such a way that it remains level. You might set it in a small paper box in the desired position. You might also permanently attach the ends of the coil to some form of connecting posts. This will help you make faster and tighter connections when you use the meter. Do not use iron or steel nails or other iron hardware in your construction. Can you guess why?

For best results when using the meter, keep it far away from all magnets, iron, and wires carrying electricity. Let the needle point north, then turn the meter so that the turns of wire forming the coil are parallel to the compass needle.

Test a fresh dry cell first. The meter should show a strong reaction. The needle may swing around wildly. Then test an old cell which is too weak to ring a bell. The compass needle will probably still show a movement.

Now for the real test of your workmanship. Is your meter sensitive enough to detect the feeble current produced when a copper plate and a zinc plate are stuck into a lemon? (See illus., p. 48.) From an old dry cell casing, cut a 1-inch by 2-inch strip of zinc. Obtain a piece of copper which is cut in the same way. Attach a 10-inch length of wire to each strip of metal. Do this by bending about ½-inch of one end of the metal over the wire and clinching it tightly with a hammer or pliers. Clean the metal plates with sandpaper.

Set both metals into a lemon in deep slits about ½

inch apart. Connect the wires to the meter. Your meter needle should move in one direction. Reverse the wires connected to the meter. The needle should move in the opposite direction.

The electricity is produced because of the chemical action of the acid in the lemon upon the two different metals. Try using other pairs of different metals for your "lemon cell." Will a penny or a dime work?

See what happens when two similar plates are used. This is your *control.* This should show you whether different metals are required in making a "lemon cell."

Follow carefully the instructions for using the meter. Here are more suggestions for increasing the sensitivity. See what happens when more turns of wire are used. Up to a certain point, as the magnetic field gets stronger, there should be an improvement. Remember however, that it is possible to add so much wire that it offers resistance to the feeble current.

Try to get the turns *closer* to the compass needle. That is why it is important to use "magnet" wire. This is a thin wire, having thin insulation, so that many turns can be wound in a small space.

How can magnetic lines of force be preserved?

There is an invisible force around a magnet. You can feel this mysterious pull when you bring a piece of iron in the neighborhood of a magnet.

You can also see the direction and strength of the magnetic force around a magnet by doing the following:

Cover a strong bar magnet with a flat sheet of smooth cardboard or a pane of glass. From a height of about 8 inches, sprinkle some fine, dry, iron filings on the cardboard or glass. The filings will arrange themselves in a definite pattern of lines called magnetic LINES OF FORCE.

BAR MAGNET WITH LINES OF FORCE

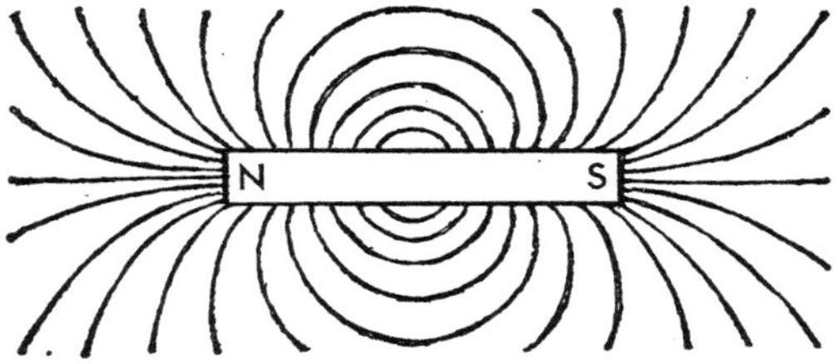

By studying these lines of force you can see the magnetic fields around various types of magnets. You can also place two or more magnets near each other, and see what happens to lines of force when unlike poles attract each other, or when similar poles repel.

Your research problem is not only to create these lines of force, but to make permanent records of them so that they can be shown and discussed at some future time.

The most obvious method is to sketch the interesting patterns which are formed. Or, if you have a camera which allows you to make close-ups, you can photograph the various arrangements of the lines of force.

Another method is to place a sheet of heavy wax paper over the cardboard or glass which is over the magnet.

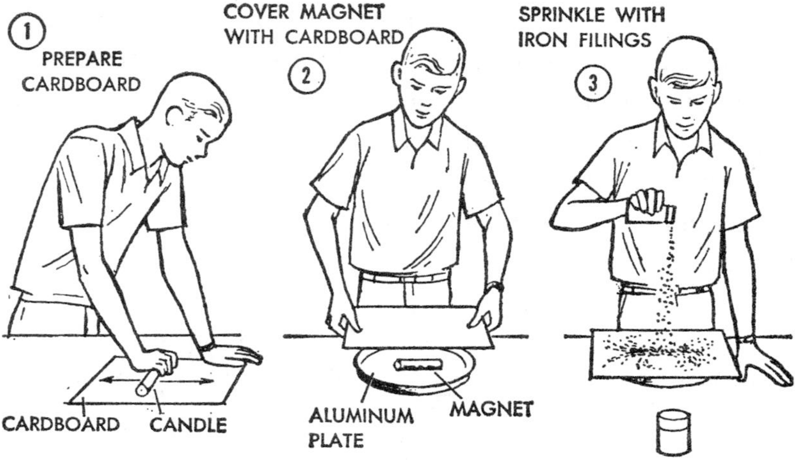

Sprinkle the iron filings to form the lines of force pattern, then heat up the wax paper in some safe manner. A heat lamp is ideal for this purpose. Some of the iron filings will adhere to the softened wax forming a permanent record.

If this set-up is done in an aluminum baking pan, it can be lifted without disturbing the filings and set over a hot radiator. However, more heat can be obtained by placing the pan with its contents in a warm (*not hot*) oven until the wax sets. Remove carefully and cool.

A more suitable and heavier wax layer can be made by rubbing the broad side of a white wax candle on a sheet of paper. Brush off the excess particles of wax with your hand. Use this prepared paper as you would the regular wax paper.

Excellent results can be obtained with blueprint paper.

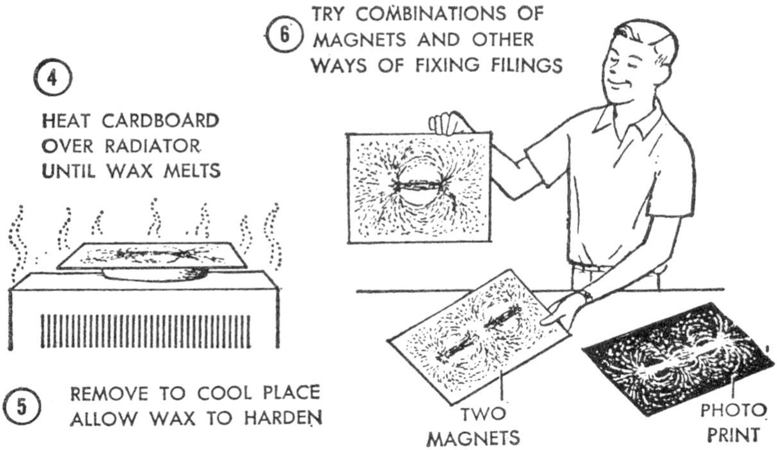
4
HEAT CARDBOARD
OVER RADIATOR
UNTIL WAX MELTS
5
REMOVE TO COOL PLACE
ALLOW WAX TO HARDEN
6
TRY COMBINATIONS OF
MAGNETS AND OTHER
WAYS OF FIXING FILINGS
TWO
MAGNETS
PHOTO
PRINT

Use it with the sensitive side face up in a shaded room. After the iron filings are in place expose the paper by bringing the entire set-up into the sun or by shining an electric light on it. Expose until the background turns gray. Remove the filings and develop the paper by placing it in water. The finished print has white lines of force against a blue background.

By far, the best prints are made with photography paper, which is used in the same manner as blueprint paper. However, a chemical developer and fixer are needed. Team up with a friend who can show you how to do this. Remember to keep any trace of iron filings out of the developer.

Another suggestion you might investigate is spraying the lines of force on the cardboard with a clear lacquer. This will make the filings stick to the paper.

The spray cans are sold in paint and hardware stores. Spray guns and spray bottles can also be used, if you wish to buy the lacquer separately. Try shellac too.

Can you think of a way of rusting the iron filings so that they leave a permanent stain on ordinary paper? Try covering the paper while the filings are on it with an upturned carton. Make the air moist by placing several open jars of water under the carton.

Still another way might be to place a sheet of carbon paper face up over the filings. Now place a sheet of white paper over the carbon paper. Will rubbing the top paper produce a carbon impression of the projecting bits of iron filings?

Hang the finished "portraits" in your room. They will serve as good conversation pieces, especially if you have several puzzlers among them.

Can you make a rain alarm?

In your experimental work with electricity you have probably made some types of burglar alarms and perhaps even fire alarms. Do you think that you can now do some technical research, and invent a foolproof and practical device which will cause a bell to ring when it starts to rain?

Give it some thought. If you are successful you will be hailed as an electrical genius by your family and friends. This world has long been waiting for such a useful gadget which will give someone at home an unmistakable warning to close the windows as the first drops fall. No more rain-soaked curtains, furniture, floors, or ceilings!

The heart of the alarm circuit is some kind of electrical switch whose contacts are prevented from touching each other by some kind of water-soluble substance. When the rain dissolves this material, then the spring switch closes, the contacts touch, the circuit is completed, and the bell rings!

One suggestion is to use a spring-type clothes pin which is found in almost every home. The clip is made of wood or plastic and is therefore insulated. Wrap several turns of bare copper wire around each gripping tip of the clip. Twist the wire so that it cannot loosen.

Separate these electrical contacts by means of an aspirin tablet which is inserted between them. To test, place some drops of water on the pill. It should fall apart in a few seconds, drop out of the way and allow the switch to make a positive contact.

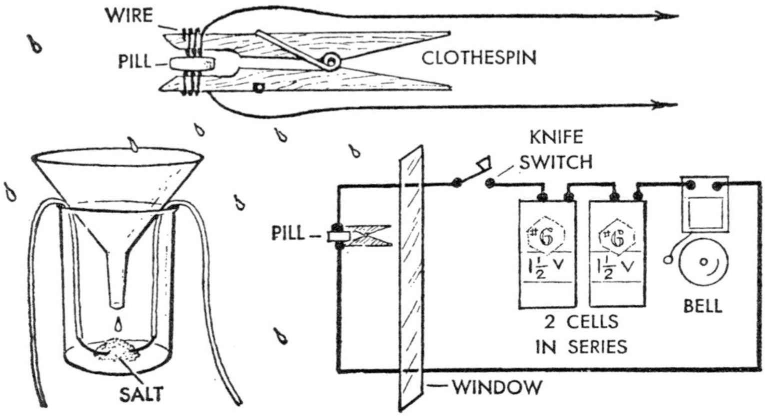

If it does not do so, then you must search for a suitable pill. Some aspirins disintegrate faster than others. Try other types of pills too. Some tablets, such as Alka-Seltzer, bubble and break up as soon as water strikes them. Some saccharin tablets do the same. They are called "effervescent," and are very small and extremely inexpensive. You might also try small bits of lump sugar and other highly soluble materials you can find in your pantry or medicine chest.

Can you devise a completely different kind of switch which will be activated when something gets wet? For example, some type of homemade spring switch which is held open by a narrow strip of paper. When the paper gets wet it breaks and the switch closes. In this case, the contacts can be housed in a dustproof container. This is an advantage.

Another idea is to have two copper contacts, very close, but not touching each other, in a small glass or plastic dish. Cover these two copper pieces with table salt. Dry salt is not a conductor of electricity, but when it gets wet the salt water is able to complete the circuit between the contact strips. You will probably need a funnel arrangement whereby a small amount of rain can be funneled into the small dish. This will quickly raise the water level and dissolve the salt.

In all your experimentation remember that the rain alarm must go on in seconds. Nobody needs a rain warning which works ten minutes after a rain starts. Make a chart showing the "batting average" of your rain alarm.

In other words, indicate when it goes on successfully.

Notice that in the illustration of the complete hookup that two dry cells are used instead of just one cell. This is necessary, since the contacts may become tarnished after being exposed to the air. A stronger voltage insures a more reliable circuit.

Do not forget to include in the circuit a knife switch, which can be left open when no one is home or at a time when you do not want the alarm to go off.

MAGNETISM AND ELECTRICITY—
More to find out:

How is an electric cell made?

You will need a copper penny and an iron washer which have both been cleaned or brightened with steel wool. Also, a 1-inch square of white blotting paper which has been soaked in vinegar. Place the wet blotting paper between the washer and the penny.

To detect the feeble current produced by this homemade cell, you need the electric meter you made when you coiled wire around a compass. See page 47.

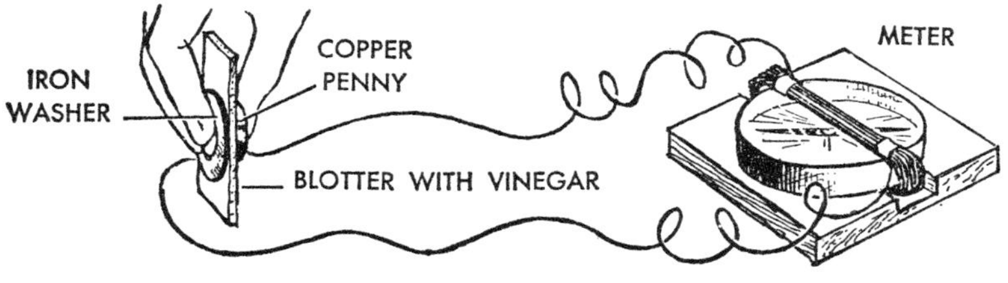

Touch one wire of your meter to the penny and the other wire to the iron washer. Squeeze the wires between your fingers to improve your "electrical sandwich."

Reverse the wires and the compass needle will go in the opposite direction. Try salt water or lemon juice instead of vinegar. Make a cell using two different metals. Will a cell work if two similar metals are used? Will larger plates of metal produce more current? Does the current get weaker or stronger when you put many "sandwiches" in series? That is, have an arrangement of copper–blotter–iron–blotter–copper–blotter–iron, etc.

Can you make an electric generator?

Here is another experiment with which you can test the sensitivity of your electric meter described on page 47. Read again the instructions for using it.

Wind a 1½-inch coil of about fifty turns of thin insulated wire. Tie or tape it to prevent unraveling. Attach the ends of the coil to your meter. *Make the connecting wires about 3 or 4 feet long!*

Quickly insert the end of a strong magnet into the generator coil you just wound. See how the compass needle moves in one direction and then returns to its original position. You may have difficulty in seeing the meter which is over 3 feet away. In that case, have a friend move the magnet in the generator coil.

Observe that when the magnet is quickly moved from the coil, that the needle moves in the opposite direction. This is where alternating current (AC) gets its name. Try holding the magnet still, and moving the coil over it. Current is produced this way, too. But, notice that no current flows when there is no movement.

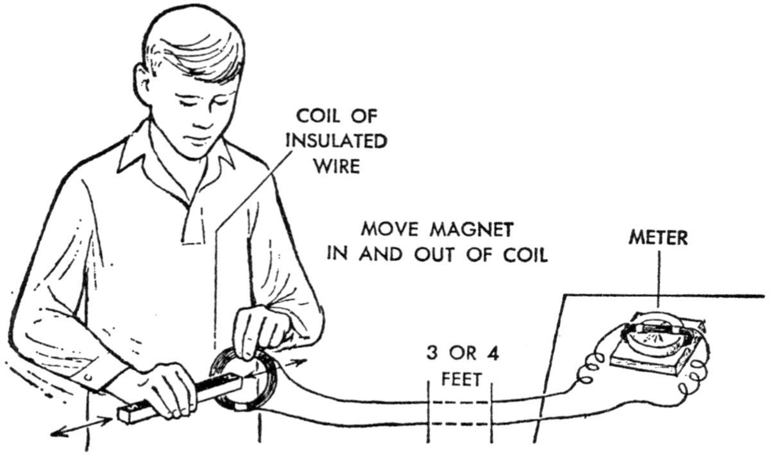

You are demonstrating Faraday's Principle which states that an electric current is produced in a wire when it cuts a magnetic field. The electricity coming into your home is generated this way.

You must test and make certain that your generator is far enough away from the meter so that the movements of the magnet do not affect the compass. Try to make the generator stronger by having more turns in the coil, using a stronger magnet and making faster movements with the magnet.

Does a magnet pick up sand?

The next time you go to a beach bring along a magnet. Try to pick up some dark sand with it. It may surprise you to find that certain sands, especially the black varieties, are magnetic. The grains consist of fine particles of an iron mineral called magnetite.

How is a storage cell put together?

Get two clean strips of lead about 2 inches wide and 5 inches long. Bend over about ½ inch of the top of each strip and hammer the crease tightly over the bared end of a 15-inch wire. Place the two lead plates close together, but not touching each other. This can be done by placing several wooden match sticks between them. Wrap rubber bands around this assembly to keep it intact. Be sure the rubber bands do not cause the soft plates to touch.

Set this into a glass tumbler or jar containing one heaping tablespoonful of bicarbonate of soda dissolved in hot water. To charge the cell, connect one wire to one terminal of a 6-volt battery and the remaining wire to the other battery terminal. Your source of current may also be several dry cells connected in series.

Allow the storage cell to charge for five minutes. Then disconnect the wires from the dry cells and connect the storage cell wires to a small flashlight bulb which should light up. A bell can be made to ring also.

Experiment by trying different sizes of lead plates

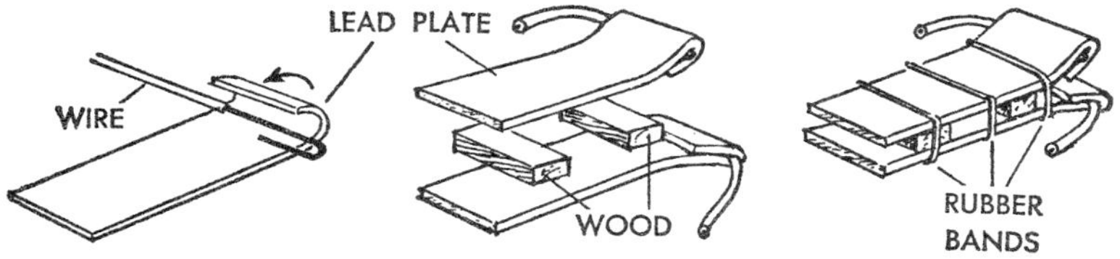

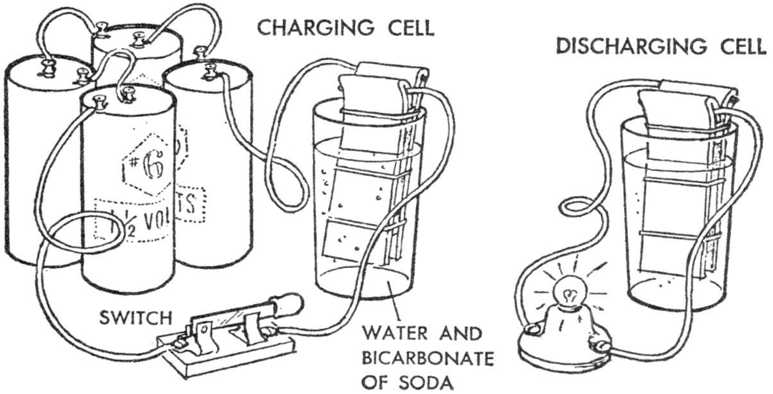

and also by varying the distance between them. See what happens when ordinary tap water is used instead of the solution of bicarbonate of soda. Try different charging times, and weaker or stronger strengths of the solution. How long does a charge last? Make charts or graphs of these relationships.

When charging, the electricity causes certain chemical substances to be formed on the surfaces of the lead plates. When discharging, the activity of these chemicals produces an electric current. The storage battery used in your car has sulfuric acid in it instead of the bicarbonate of soda. It also consists of many storage cells hooked up in series.

Can a dry cell be revived?

A dry cell cannot usually be recharged. When it no longer rings a bell you have to discard it. However, a dry cell is really not completely dry inside, and sometimes the moist chemicals inside get a little dry and the cell stops working. Such a cell can often be revived.

Remove the paper cover and punch several nail holes in the zinc bottom. Set the cell into a jar of salt water about 4 inches deep for five to ten minutes. See if you have restored it by connecting it to a bell or bulb.

If you are successful, pour off the salt water. Use the dry cell while it is in the jar, to prevent dripping. Or perhaps you can think of some way to prevent dripping by sealing the holes.

Can you light a fluorescent lamp in your hands?

Remove a fluorescent lighting tube from its sockets and take it into a darkened room. Rub it briskly on a woolen garment and it will light up! As with all experiments involving static electricity, this one works best on a cold, dry day.

Rubbing gives the glass an electrical charge by removing electrons. Movements of these negative electrical particles produce the current which lights the lamp in this magical manner. See what happens when different materials are rubbed instead of wool. Try silk, nylon, hair, plastic, rubber, leather. See whether rubbing an ordinary tungsten electric light will also light it up.

WEATHER

How far can you see today?

Have you ever observed distant landmarks from an upstairs window at home, in your school, or from some other high place? You may have discovered that on certain days you can see further than at other times.

The greatest distance along the ground at which objects can be seen and identified is called VISIBILITY by weather men. It is useful information to many people, especially to airplane pilots who are about to take off or land.

Before they visit the Empire State building observatory sightseers in New York City may wish to know how many miles they can see on that day. Visitors to mountain lookouts may also wish to know the visibility in advance. Sailing enthusiasts, too, are eager to know the furthest distance that they can still see the shore.

In fact, so many people are interested in the visibility, that radio and television commentators usually include this in their daily weather reports.

Visibility is affected by humidity, dust, smoke, and other substances in the air which reflect light and cause a haze. The presence of clouds and the hour of day are also reasons for poor visibility.

You can make a visibility chart like the one shown in

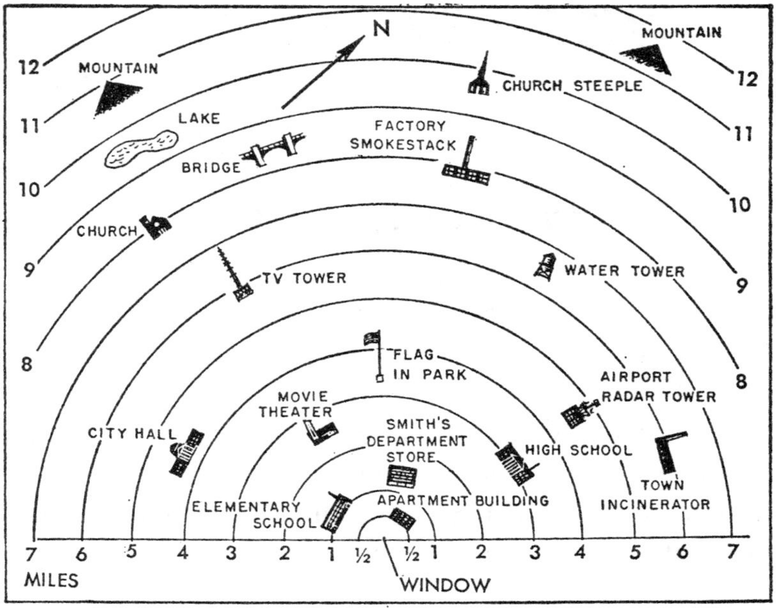

the illustration by finding the distance certain landmarks are from your observation post. There are many ways to do this.

In many cities and towns good street maps are issued by gasoline companies or sold at newsstands and stationery stores. You can also obtain copies in the library. The city Surveying, Highway, or Real Estate departments may allow you to see their maps. Each map has a scale in miles per inch.

Perhaps some driver of the family automobile can help you measure the miles by using the part of the speedometer called the ODOMETER. This records miles and tenths of miles. Remember however, that you want the straight line distance between the landmark and your eye. You can often get this measurement accurately by driving up a neighboring street which runs in the same direction, even though it is not in the exact line of sight.

Many cities have streets which have blocks of equal length. By multiplying the number of blocks by the size of each block, you can obtain the total distance. Do not forget to include the paved distance between the streets.

Use a compass to draw circles or arcs on the chart, using your window as a center. Your furthermost landmark should be the one you can see on the clearest day. Your nearest one should be 50 feet away. Use this for a foggy day. Since this distance may not fit the scale of the regular chart, you may wish to make a special enlarged chart for closer distances.

Is the visibility the same in all directions? You can find out by making other visibility charts from windows looking in opposite directions.

To demonstrate the difficulty in estimating distances, ask your friends to guess the miles to certain landmarks. They will be amazed at their errors.

Try to verify your observations of the visibility with the figures given by the weather reporters. Remember that visibility changes with the time and location, so compare notes carefully.

Does a strong wind coming from a certain direction

usually increase visibility? What causes sudden changes in visibility? Is there a close relationship between accurate weather prediction and visibility? Keep careful records for several weeks on a chart.

Is a cobalt chloride moisture indicator reliable?

Almost everyone is familiar with this humidity detective. It consists of a piece of paper or cloth which has been soaked in a solution of cobalt chloride and then dried.

When thoroughly dry the material is *blue.* It turns *pink* when there is a considerable amount of humidity (invisible water vapor) in the air. For this reason it is used to predict rain. While many people boast of good results, some observers say that this device is a meaningless toy, since it does not turn pink *far enough in advance* of a rain to be of any value.

You can obtain some cobalt chloride from your science teacher. Most chemistry sets also have it. Dissolve ½ teaspoon of the crystals in ½ glass of water. Place into this solution different kinds of cloth, white blotting paper, paper toweling, or any other paper you wish to use. Allow to soak for several minutes. Remove the materials from the liquid and dry them.

1. How closely does your indicator check with the actual weather conditions?

Keep a chart for a few weeks. Record the date and hour, color of the indicator, weather condition, and the relative humidity given by the Weather Bureau. Of course, if you have your own hygrometer (humidity tester) it would be better since you can place it next to the cobalt chloride devices which you are testing.

DATE AND HOUR	LOCATION OF INDICATOR	COLOR OF INDICATOR	WEATHER	RELATIVE HUMIDITY
12/6 8 A.M.	OUTSIDE	#4 BLUE	DRY	52%

2. Does it matter where you place the indicator?

Is the humidity the same all over the house? Is the humidity inside a house the same as outside?

Place one piece of cloth or paper prepared with cobalt chloride inside the room away from a window. Tape a similar indicator on the inside of the window frame. Now tape an additional piece of the same material outside the window, but shielded from the direct rain.

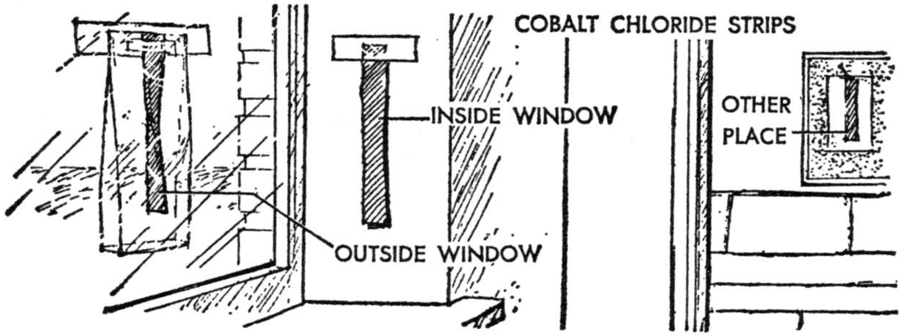

This can be done by taping your indicator to the window by means of adhesive plaster or cellophane tape and by taping a flap of thin plastic over it. Be sure to allow an opening on the bottom for the outside air to reach the indicator.

Another method is to tape the paper or cloth to the

inside of an open plastic bag. Tape the bag to the window with the opening down.

Label each indicator. Keep records similar to those in the first experiment.

3. Which material makes the best indicator?

Soak various materials in the same solution for the same length of time. Try cotton, wool, rayon, blotting paper, writing paper, and other substances.

Place all the indicators *in the same place*. Keep careful records to find out which is the most sensitive and accurate.

4. Does a stronger solution of cobalt chloride make the indicator more sensitive?

Make a stronger solution of cobalt chloride by adding more crystals to the same amount of water that you used before; perhaps 1 teaspoonful of crystals to ½ glass of water.

This time use only one kind of material for the indicator since this experiment has to do with the strength of the material and not with the type of materials used.

You may also wish to try the effect of a weaker solution. Make one by adding more water to the original solution.

Label your samples. Carefully record in your notebook the strengths of the solutions used for each. Place all the indicators in the *same* part of the house when doing this experiment.

5. Can you think of more ways of testing or improving this humidity indicator?

What is the wind direction in the upper air?

If you have any kind of weather eye you will no doubt have noticed that the direction of the wind on the earth's surface is frequently different from the wind high up in the air. You know this because you observed the direction of the clouds.

Seeing how the clouds are drifting is one of the methods used by the professional weather man for determining the wind direction high above the earth. He may also send up a special helium-filled balloon and follow its path with a telescope or by radar. Other intricate electronic devices have been developed to get information about the winds high above the ground.

There is a good reason why this information is very valuable in weather forecasting. The large, broad winds which blow the clouds, give a more accurate picture than the surface winds give, of the huge forces which are moving masses of air across the country.

Surface winds often give false information because of local obstructions and disturbing conditions, such as cities and towns, forests, plains, lakes, and hills.

You can make an accurate instrument used by meteorologists for observing the cloud direction, without having to crane your neck. It is called a NEPHOSCOPE (NEF-a-scope), and is simply a circular mirror held horizontally so that it reflects the clouds to your eyes. The edges of the mirror are marked with the compass points to indicate the direction the clouds are moving.

Look at the illustration. Try to duplicate the set-up by any means you have available. In the center of the mirror place a dot. Do this by using gummed paper, nail polish, India ink, or some other marker. You can use the same material for indicating the four compass

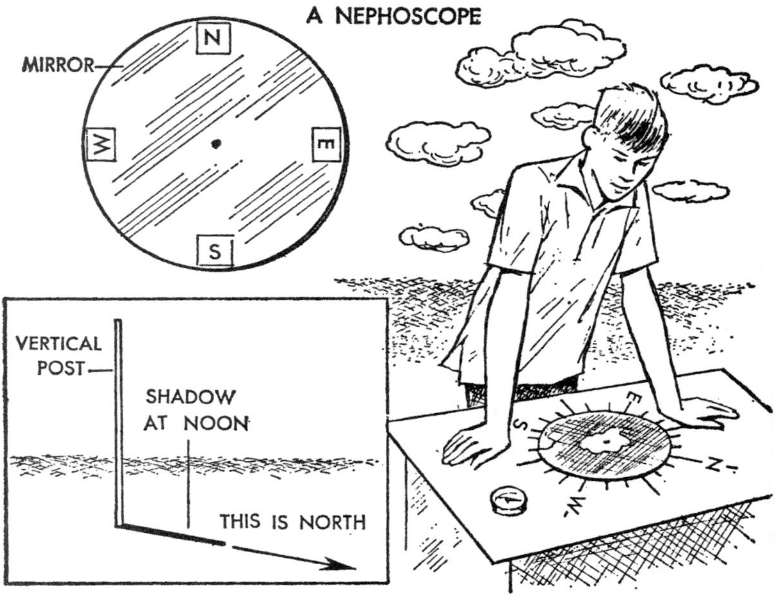

points. You may wish to place the mirror upon a heavy cardboard or piece of wood. If you can get only a square mirror, paste a paper with a cut out, circular hole over it.

To use the nephoscope, set it on a level place and turn it so that the "N" mark faces north. You can get help in finding north by using a local map obtained at a gasoline service station. If you use a compass, remember that there may be a slight error because the needle points to the magnetic north and not to the geographic or true north.

The best way to find north is to notice the direction of the shadow cast by any vertical post exactly at noon. The shadow line will be going toward the north, *away* from the post.

With the nephoscope properly positioned, look into the mirror and observe a cloud, or part of a cloud which *crosses* the center dot. Hold your head in one position until the cloud drifts to the edge of the mirror. Now read the direction from the markers.

This is the direction *toward* which the wind is blowing. Since a wind is named for the direction from which it comes, the type of wind is opposite to where the cloud left the mirror's edge. In other words, it is an east wind which blows the clouds to the west.

Keep records such as the sample below to learn how much difference in direction there is between surface winds and upper-level winds. At the same time you can record any other helpful information.

DATE	DIRECTION OF WINDS		TYPE OF CLOUDS	WEATHER CONDITIONS	ETC.
	SURFACE WIND	UPPER WIND			
MARCH 4 2 P.M.	S E	E Note: Some clouds were being moved by N E wind	CIRRUS	INCREASINGLY CLOUDY	50° F

What is the best way to catch snowflakes?

Snowflake catching is like panning for gold. You have to look for the beautiful, six-sided, star-shaped crystals as they fall. Snowflakes may crush, melt, or cling together. In fact, the crystals you are looking for may not have been formed in the first place. You have to wait for the right kind of snowfall. For example, a wet snow is rarely good.

Another difficulty is that snowflakes melt almost as soon as they fall. Try different ways of catching them. One way is to stand in a partly sheltered place and hold out a flat sheet of felt, velvet, or any similar material. Look closely at one snowflake and draw a picture of it immediately. If you have the proper camera you can photograph the crystal. After you draw many of them you will see that no two crystals are ever exactly alike. Does it help to chill the cloth in a freezer first, or to keep the cloth over some ice cubes or dry ice?

How accurate are thermometers?

Check several inexpensive thermometers against a good thermometer whose accuracy you are sure of. It is important to check different points on the scale. A poor thermometer may be accurate at 70 degrees, but wrong at 28 degrees. Test the thermometers under the identical conditions. Keep them close together. Allow time for temperature adjustment during each reading.

You can also check fixed points directly, such as the temperature of melting ice and the boiling point of water. Do this by placing the thermometer into crushed ice and water, or in boiling water.

Does a thermometer, even a good one, give the same reading when it is lying down, on its side or upside down? Try this with several thermometers at various temperatures. The results will surprise you.

Can slant of rain tell you the speed of the wind?

Measure the angle made by rain which is falling in an open area. As a reference point for your protractor, use any vertical object such as the side of a building or a pole. A handy plumb line is shown below.

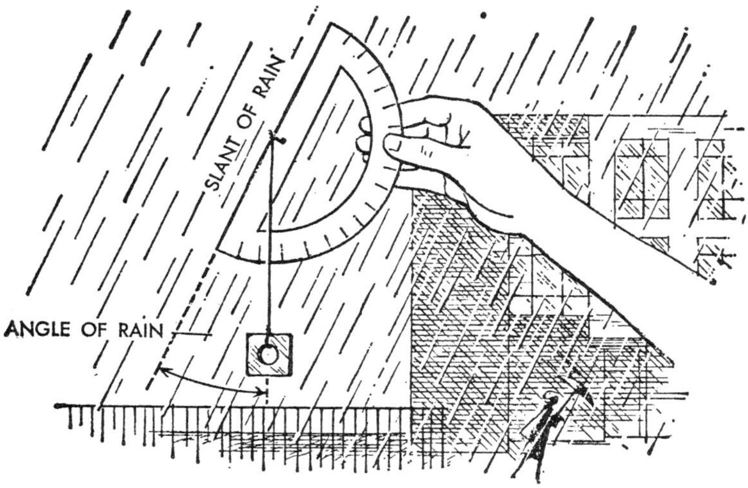

Use your ingenuity to find out the local wind speed. Are the wind speeds given by the radio weather reports usually accurate for your neighborhood? Is there an anemometer near you which gives the wind speed? Compare conditions with the Beaufort wind-speed scale. This is found in encyclopedias and books on weather.

Is it important to record the size of the rain drops? Should you indicate whether it is a drizzle or heavy downpouring? After your research, will you be able to report the wind speed just by seeing the slant of the rain? Is there any connection between the direction of the wind and the slant of the rain?

How does the temperature of a lake change?

The temperature of a big body of water does not have large or rapid changes. On the other hand, the temperature of the air may vary greatly during the day. You can show this on a graph which contains data gathered by you. It will have two lines; one showing the water temperature, and the other line the air temperature.

Carefully take the temperature of the water of a lake, pond, reservoir, river, or ocean. Record the temperature of the air at the same time. Do this at regular intervals during the day for several days or longer, if possible.

Can you make a huge hygrometer?

A hygrometer is an instrument which measures humidity in the air. Many professional models contain strands of human hair which lengthen when the moisture increases in the air. The hairs shorten when the moisture decreases.

Stretch a rope outdoors between two posts, or other solid supports about 10 feet apart. Attach a weight to the middle of the line. Erect a post in the ground containing a vertical yardstick and placed alongside the weight. This will show the rise and fall with the humidity.

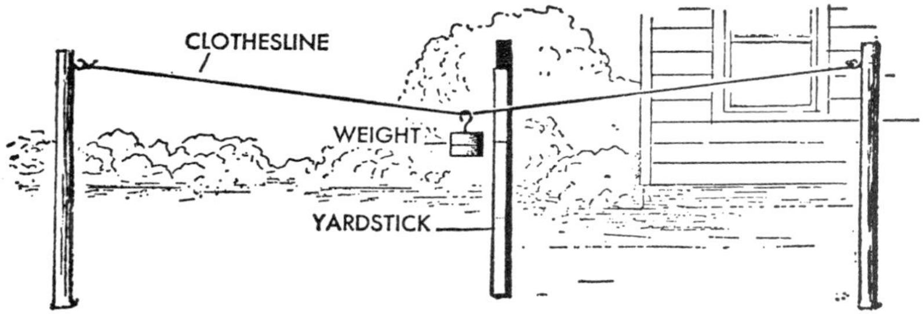

Allow several days for the rope to get adjusted. Then keep careful records of the humidity as indicated by the weather reports. At the same time, record the position of the weight. Does the rope behave like a human hair? Try using different kinds of ropes, lighter and heavier weights, shorter and longer lengths of rope. Does an indoor rope hygrometer behave in the same way as an outdoor one?

Should your radiators be painted black?

It may not look decorative but your home would be heated better if your radiators were black. You can easily convince any skeptics by the following demonstration:

Get an empty, shiny tin can about 3 or 4 inches in diameter, and about 5 or 6 inches tall. Wash off any grease. Paint half of the can from top to bottom, inside and outside, with a dull black paint or India ink. Smoking it over a candle is also effective. Place a 1-inch long candle exactly in the center of the circular design on the bottom of the can.

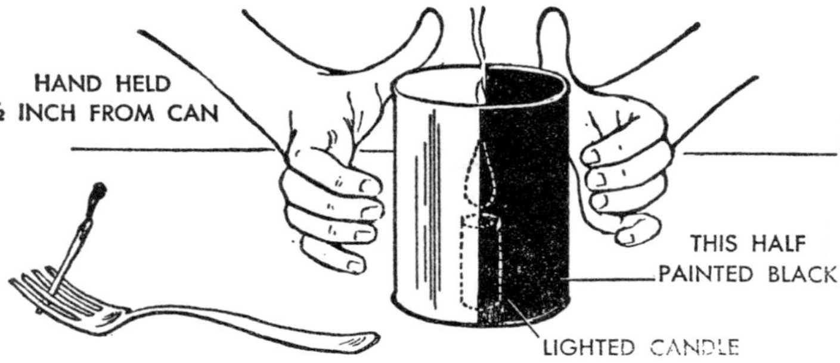

Light the candle. This is easily done if the match is stuck between two prongs of a fork. Keep away from drafts. Cup both your hands around the black and the shiny sides of the can. Keep them about ½ inch from the metal.

You will find that the black side feels warmer. You may also touch the different sides, especially near the bottom. Experiment with different sizes of cans and candles. Since black is a good radiator of heat, it should also lose its heat easier. Test whether it cools off faster than the undarkened section.

WATER

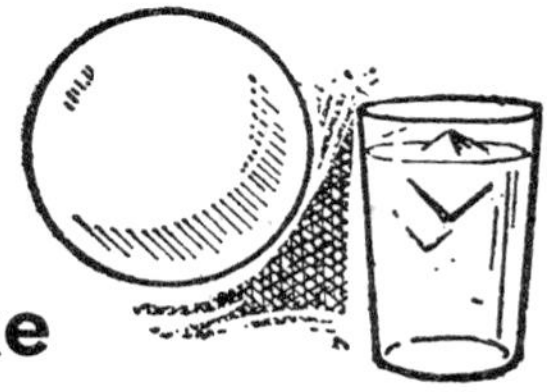

How large can you make soap bubbles?

By using your own formula you can produce huge soap bubbles which will surprise you. There are various facts and techniques to be learned however, before you become a bubble expert.

Do you know why a soap bubble is spherical? It is due to the fact that a liquid surface behaves as though it is an elastic film. It is constantly pulling itself in, so that it makes as small a surface as possible. This "pulling-in" force is known as SURFACE TENSION. Scientists believe that it is caused by the attraction that molecules of the same substance have for each other. This is called COHESION (koh-HEE-zhun). A bubble breaks when it touches something, because the attraction it has for other molecules (ADHESION) is greater than its cohesion.

A soap bubble is a small amount of soapy liquid spread out so that it forms a closed surface. This surface is constantly and evenly pulling itself tight. The surface becomes spherical because a sphere is the smallest surface that any volume can have. The spherical shape also means that pressures are equal on all sides.

There are several ways to produce bubbles. You can use a bubble pipe which is dipped into the bubble liquid and then blown upward or downward. A slight twist of the wrist releases the bubble at the best moment. Do this out-of-doors, or in a garage.

Another method, giving you more control, is to use the wire loop shown in the illustration. Obtain some

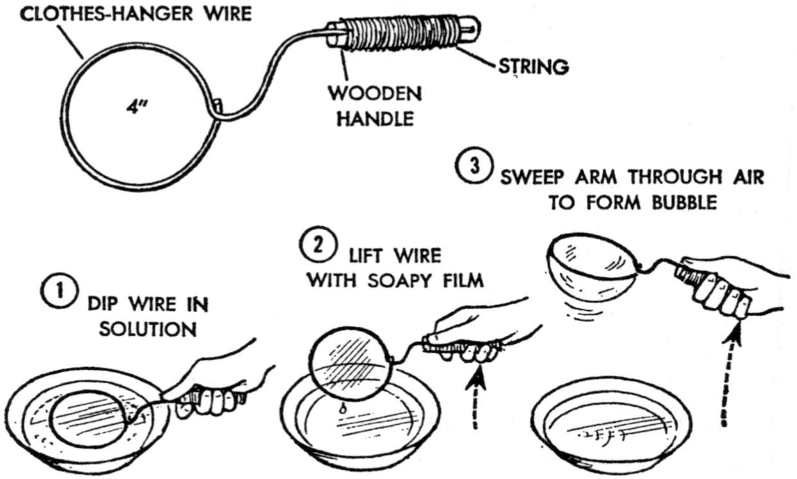
CLOTHES-HANGER WIRE
4"
STRING
WOODEN HANDLE
3 SWEEP ARM THROUGH AIR TO FORM BUBBLE
2 LIFT WIRE WITH SOAPY FILM
1 DIP WIRE IN SOLUTION

medium-stiff wire used for clothes hangers. Make a 4-inch loop at one end by winding the wire around a jar. If you wish, you can make a handle by attaching the other end of the wire to a round piece of wood by means of many turns of string.

When using the wire bubble maker, dip the ring into the soap bubble liquid. Remove it horizontally and slowly. It should have a soapy film in it. Waft it upward, not too slowly and not too quickly. At the correct speed the film will form a bubble and separate from the loop.

Most of the experimental work necessary to produce

large soap bubbles lies in developing your formula for the best bubble liquid.

Use the following ingredients as a starter:

1 part liquid detergent (used for dishwashing)
1 part glycerine
5 parts clean water

Use a soup dish for a container. Liquid soap or soap powder may be substituted for the dishwashing detergent. Vary the different materials in the above formula *one at a time* and observe the effects. Also try many different soaps found in your home. Keep records in your note book of each trial.

Add a pinch of sugar to the above formula. It should toughen the bubble. Keep adding sugar until the maximum benefit is reached. Use this amount in your future formulas. Above all, keep the bubble liquid and the bubble makers clean.

Does heating or cooling the liquid have any effect upon the size of the bubbles? Does it help to make the liquid sudsy?

Is there any improvement when you allow the soapy film in the wire ring to drain, by holding it at a slant over the dish? Does filtering the liquid through cloth give better results? Do wire loops of different sizes produce more durable bubbles? Does the thickness of the wire determine how much liquid goes into each bubble? Will a loop made in the shape of a square produce spherical bubbles?

While you are working with soap bubbles, here are more things you might like to experiment with.

Can you use a wire loop to catch a bubble which is in the air? Use a tough bubble and a *wet loop.* Can you catch a bubble between two loops and pull it into a long shape?

What happens when a bubble is wafted over a hot radiator or lamp? Can you bounce a bubble on a woolen blanket? Can you pat a bubble with a hand covered by a woolen mitten? How long can you fan a bubble between several people?

Place a cup with the opening down, into bubble liquid. Lift and there will be a film over the opening. Study the play of colors on it, especially in sunlight.

Try blowing many small bubbles at one time by making a multiple blower. Do this by attaching many drinking straws alongside each other.

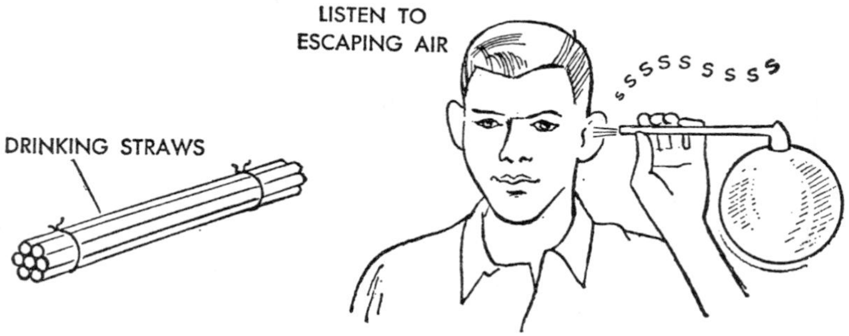

You can show that a soap bubble contracts. Blow a large bubble with a pipe whose bowl is facing downward. Do not detach the bubble. Put your ear to the stem and hear the air being forced out by the shrinking bubble. This tiny wind can also make a candle flame move.

Can water containing ice get warm?

Have you ever seen a glass of ice water standing in the hot afternoon sun? You know that the ice cubes, water, and glass are absorbing heat which is radiating from the sun. It seems natural to assume that the water must be getting warmer. Very often while there is still some ice left in a glass, a well-meaning, but unscientific hostess will say, "Let me add more ice to make your drink cooler."

But can ice water really get warm as long as there is any ice in it? To answer that question there are certain facts you should know.

You have no doubt heard of molecules of water. These are the smallest particles of water which are possible. Of course, no one has ever seen them. Scientists believe that molecules are always moving about, and in a liquid such as water, they slide over each other very easily. That is why water can be poured and takes the shape of its container.

However, when the water molecules lose their heat and are cooled, they do not move about so freely. At 32 degrees Fahrenheit, they slow down so much that they become part of a rigid, frozen material we call ice.

It takes a great deal of heat to loosen the molecules and to change ice back to water. In fact, it takes a special quantity of heat just to change *ice* at 32 degrees Fahrenheit to *water* at 32 degrees Fahrenheit—the same temperature! This is where all the apparently "lost" heat is going when the ice water is in the sun.

Theoretically, as long as there is one chunk of ice left in the glass, any heat which is absorbed will go toward melting the ice, instead of raising the temperature of the water. As you have no doubt noticed, after the last piece of ice is gone, the temperature of the water rises rapidly.

See for yourself by doing this experiment. Fill a large drinking glass with ice water containing three or four ice cubes. Set it outdoors in the sunlight. Place any household thermometer you have available into the glass. It will probably read 32 degrees Fahrenheit, or perhaps a degree or two higher.

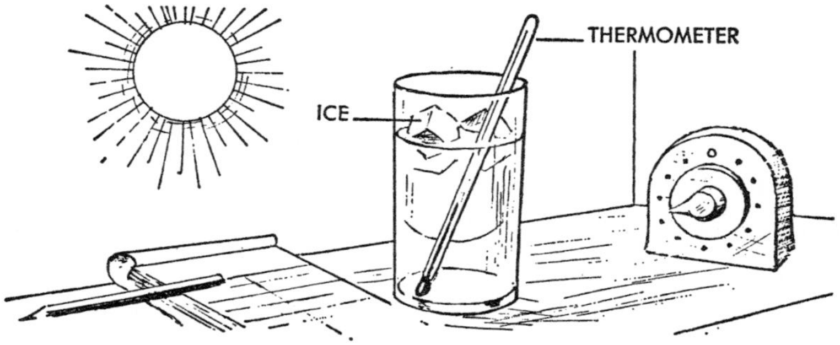

Take the temperature of the water every five minutes. Use a timer, if you have one. Constantly and carefully stir the water and ice cubes with the thermometer. You will find that more vigorous stirring will be necessary as the ice cubes become smaller.

Illustrate your results by means of a chart or graph. Your notes should indicate when the last piece of ice melted. Repeat the experiment indoors. Try smaller

TIME (IN MINUTES)	TEMPERATURE (IN DEGREES FAHRENHEIT)
START	32
5	33
10	33
15	34
20	33
25	33
30	33
35	34
40	36
45	41
50	47
55	53
60	61
75	76
90	81
AIR TEMPERATURE: 84°F	

TIME LAST PIECE OF ICE MELTED (between 35 and 40)

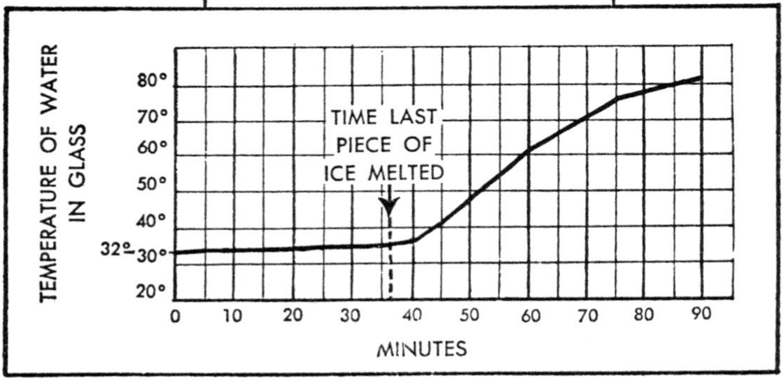

and also larger quantities of water and ice. Vary the intervals at which readings are taken. Indicate the temperature of the air.

There is another, and more spectacular way to prove that ice water cannot be warmed as long as one ice cube is left in it to demand the heat.

Put about a dozen ice cubes into a small container, preferably a glass one, used for heating water. When the temperature of the ice water drops to 32 degrees Fahrenheit, place the pot on the gas range over a very small flame. If you have an electric stove, use the lowest heat. Throughout this experiment, do not allow the thermometer to rest on the bottom of the container.

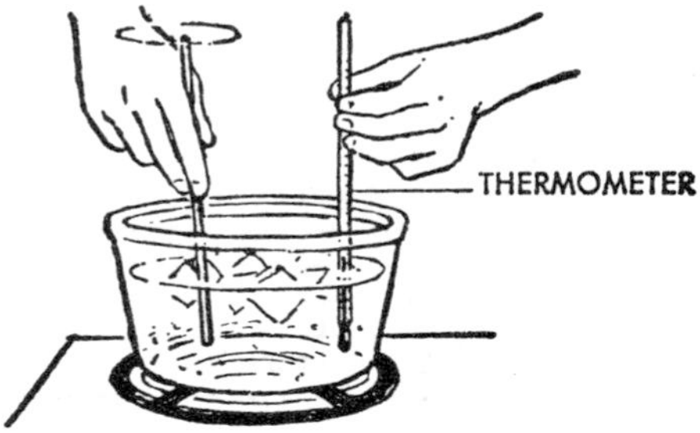

Take temperature readings as you did before, but this time at thirty-second or one-minute intervals. For this reason it is good to get an assistant. It is very important to *stir the ice water constantly*. You will find that the temperature which is maintained before the ice melts, is slightly higher than in the previous experiment using the drinking glass.

Nevertheless, your records will show you that as long as ice is present the temperature of the water is kept rather low. Then, very dramatically, when the last ice melts, the temperature rises very rapidly.

Because ice takes up much heat when it melts it is used in refrigerators. The heat is drawn from foodstuffs. Of course, when foods lose heat they get cold. This stops the growth of bacteria, and the food is preserved.

WATER—*More to find out:*

How can you make water wetter?

Cut some heavy string into ½-inch lengths and drop them into a glass of water. They will probably float on the surface. But when liquid detergent is added, the string will sink. Do the same experiment using steel wool (not soap pads), feathers, sulfur powder, fine sawdust, talcum powder. Try many other materials which usually take time to become waterlogged. Test different detergents too. You might start with two glasses of water. One should contain detergent, but not the other. The latter will be a control.

The objects float on the water because of the elastic film on the surface called surface tension. The materials have to break this tension before their surfaces get wet. The detergents, called wetting agents, decrease the surface tension of the water. Now the materials get wet, become heavy, and sink.

How big is a drop?

Kitchen recipes, medications, and chemical experiments frequently call for a certain number of drops. But are drops always the same size? Think of many ways of testing what causes variation in drops.

Is the size determined by the nature and thickness of the fluid, the opening of the dropper, the angle at which the dropper is held, temperature, or air pressure? Get a very small, narrow vial. Count the number of drops needed to fill this vial under certain conditions. Of course, the more drops you need, the smaller will be the size of each drop.

Avoid varying the conditions other than your experimental one. For example, you may wish to find the effect of temperature upon the size of drops. You must therefore use the same dropper, held at the same angle. The liquid you use must be the same, as well as all the other conditions. Only the temperature is changed in your trials.

Can you see under water?

Have you ever been in a glass-bottom boat? There is a new world of interesting things to see under the water. Because of surface ripples and reflections from the sky this world is hidden from most of us.

It should be very rewarding to experiment with different kinds of homemade waterscopes. You can have one immediately, just by using a transparent plastic bowl or casserole with a flat bottom. Try other devices too. Avoid using glass because of the problem of breakage.

Use the waterscope by merely placing the bottom below the surface of some shallow water and peering into it. It works best when the water is clear and when there is sunlight overhead. Try not to stir up mud with your feet.

Take a partner along and go hunting with the device at the beach, in shallow brooks, and the edges of ponds and lakes.

Read some library books about the fascinating animals and plants which abound under water. You will also need a kitchen strainer, fish net, and some plastic collection jars.

Does melting ice raise the water level?

This experiment seems to defy logic, but seeing is believing! Place four or five ice cubes in a large glass. Add water until the level is almost up to the brim. Watch the level as the ice melts. It does not change.

Everybody expects that when the large volume of ice melts it will form a similar volume of water. The level should therefore rise and overflow the glass. However, as you probably already know, water expands when it freezes. Therefore, when it melts it contracts. In fact, it shrinks just enough to occupy the volume of the water displaced by the ice cubes. The level remains the same!

Can you make a hydrometer?

An object which floats up to a certain depth in water, will sink deeper when it is placed in a liquid lighter than water. It will not sink as deeply into a heavier liquid. An instrument which measures the heaviness of a liquid by using the above principle is called a HYDROMETER (high-DROM-i-ter).

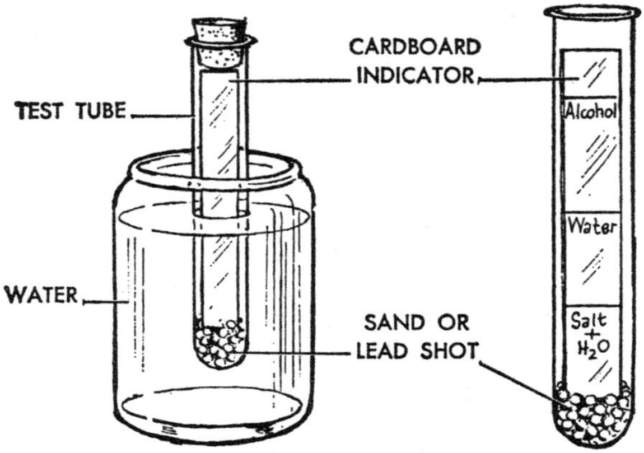

You can make one by obtaining a long test tube or any open tube which can be corked at both ends. Weight the bottom with sand or lead shot from toy air rifles. This enables the tube to remain vertical in the liquid. Insert a strip of cardboard for indicating the names of the different liquids you will test.

Place the hydrometer into a deep container of water. See how far it sinks. Mark this level on the cardboard indicator in the tube by writing, WATER. Place the hydrometer into other liquids and mark the indicator at the proper levels. Try strong salt water, rubbing alcohol, milk, cream, kerosene, and many other household liquids. Wash the hydrometer carefully each time so that you do not contaminate the different liquids, especially foods.

Are there empty spaces in water?

Suppose you had a large drinking glass of water filled right up to the top? Do you think that it is possible now to dissolve several tablespoons of salt in it without causing the liquid to overflow?

Try it and see. Place the filled glass in a saucer. Slowly sprinkle salt into the glass from a salt shaker, stirring constantly with a thin wire. Do not shake the table. Keep adding salt and stirring until overflowing occurs.

Where does this extraordinary amount of salt go? Chemists believe that there are spaces between the molecules of water into which the dissolved solids can fit. Of course, as you continue adding salt, the level in the glass rises slightly. Also, up to a certain point, the bulging surface is prevented from overflowing by surface tension.

Try this experiment with sugar or other soluble materials. Will containers with wide openings hold more material? See whether temperature makes any difference by using cold and also warmer water.

YOUR BODY

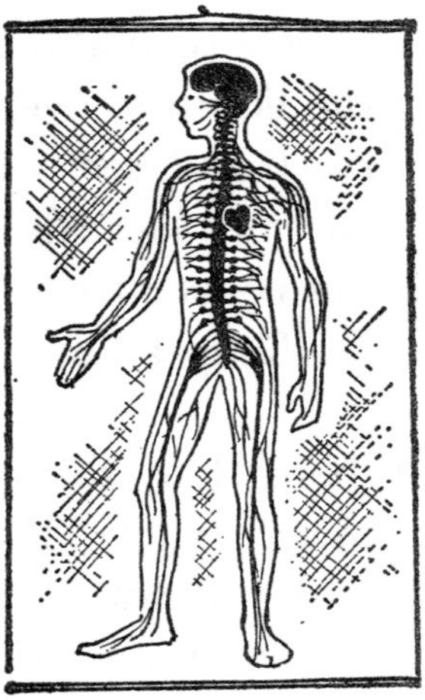

How fast is your reaction time?

Your body is alert to many things that stimulate your senses. Some examples are heat, sound, light, pain, odor, taste, and pressure. When your body responds to these in some muscular manner, it seems to do so immediately. However, a fraction of a second usually elapses between the stimulating affect and the response of your muscles. This interval is called REACTION (ree-AK-shon) TIME.

There is a wide variation in reaction time among your friends. Indeed, reaction time for the same individual may vary during the day. Fatigue, hunger, worry, sleepiness, lack of interest, or even overeating can make a person more sluggish. Remember too, that reaction time may vary for different actions. Blinking an irritated eyelid is different from pressing an automobile brake pedal during an emergency.

You can test the reaction time of your friends and classmates by several simple means. The following method gives the average reaction time of people in a group and can be done in the classroom.

Everybody, including the teacher, stands in a large circle holding hands. One person, who is also in the circle, acts as the timer. He holds in his left hand a watch with a second hand. Better still, he may be able to borrow a stop watch used in schools for examinations, science, or athletics.

All the persons are instructed to close their eyes. Timing begins at the moment that the timer squeezes the hand of the student on his right. As soon as this person feels the pressure he squeezes the hand of his neighbor. This goes on until the timer has his left hand or arm squeezed, and the stop watch is stopped, or the time noted on the wrist watch.

The average reaction time for each person is obtained by dividing the time by the number of members in the group. Do this several times until you get a consistent result. Learn whether there is a difference in reaction time taken in the morning, before and after lunch, and at the end of a school day. Does practice decrease the reaction time? To find out, do the above experiment every school day for two weeks.

Another method of testing reaction time is to drop a coin, small ball, or marble from different heights over the foot of a person facing you. He has a clear view of the object, but he does not know exactly when you will suddenly release it over his foot, which is thrust out on the floor. He is supposed to remove his foot, so that it is not struck.

Hold the object to be dropped between the thumb and forefinger of one hand. Hold a yardstick vertically with the other hand to measure the distances. Find the lowest distance the subject can avoid the falling object three times in five.

This method of testing reaction time depends upon the scientific fact that when objects fall to earth they do so at a certain rate of speed. Disregarding air resistance, for every second of free fall, the speed of an object increases 32 feet per second. It is therefore possible to calculate the time it takes for an object to fall a certain distance. The following table is based upon this principle:

REACTION TIME (IN SECONDS)	HEIGHT OF DROPPED OBJECT (IN INCHES)
.15	4
.20	8
.25	12
.30	17
.35	24
.40	31
.45	39
.50	48
.60	69

Keep reaction time records of many of your friends. Try to find answers to questions such as the following: Do those with faster reaction time have better school averages? Do older people get poor scores? Remember to use many cases before drawing conclusions.

What helps your memory?

You can assist your memory with a scientific idea which dates back to Aristotle, the Greek philosopher, over two thousand years ago. It is called association of ideas. The expression means that when thoughts are connected in some way, the later recall of one thought helps to recall the others.

He also stated that the most remembered associations are those which have either very strong similarities or very strong differences. We can also more easily recall those associations which are closer in time and space.

You can demonstrate these time-tested principles and

also learn how your mind operates. Amaze your friends and relatives by performing the following psychological experiment.

On a sheet of paper, or on a blackboard, write a column of numbers from one to twenty-five. One person, or a group is requested to give the name of a common household object for each number. Either you, or an assistant, writes the word to the right of the number as it is mentioned. Do this slowly and in regular order, starting with number one and ending with number twenty-five. As the word is mentioned and written, you may repeat it, apparently for emphasis. In a group, point to a person indicating when you are ready for the next word. You do not have to look at the list. You may even be blindfolded.

After the list is completed you will be able to recall *every* word, in the order in which it was written, without looking at the list. What is more, you will be able to name the word written after any number. You will be able to remember the words for hours, even days. Of course, your audience need not know that you do this phenomenal feat by association of ideas.

In order to be able to do this "mental magic" stunt you must first make up your own memorized list of twenty-five key words. You will use the same list every time, so you must be careful in your selection. It may sound alarming to have to learn twenty-five words, but association has made that easy for you too.

Since you already know your ABCs, you are going to

use the alphabet to help you select your key words. Once you learn it this way you will find it easy to remember.

Simply choose a good word which you can associate with the *sound* of a letter of the alphabet. In other words, ape sounds like A, bee stands for B, sea for C, December for D, etc. See the suggested list. Try to make your own personalized list.

Select words with which you can make very strong associations. For example, if your key word for A is ape, have in your mind a very special ape that you have seen in a zoo. Now, suppose your audience offers the word *chair* as number one on the list to be memorized. You can make a funny, utterly ridiculous picture in your mind of two dozen chairs in the monkey cage in the zoo; monkeys are sitting on chairs, balancing chairs, hitting each other with chairs, swinging from trapezes having chairs instead of rings, etc.

Later, when you think only of *ape*, "your number one key word," *chair* will be recalled immediately. Such is the power of association.

In your research for strong words, you will find that certain words are extremely good. You never miss with them. Try to discover why they are particularly efficient association words. Perhaps it is because these words are dramatic, ridiculously funny, definite, not confusing. Do not use two words which have closely associated or duplicate ideas. For example, do not select dog as one word and wolf as another.

Notice that only twenty-five words are used. If you select twenty-six words it might be a tip-off that you are using the alphabet to help you.

When performing this experiment, start off by keeping the key word for A in your mind. Point to a person indicating that you are ready for his word. As it is being written, you must make a strong, wildly imaginative and preferably ridiculous picture in your mind, using the two words. Repeating the given word helps fix it in your mind.

When you point to the next person, you already have in mind your key word for B. After a few times this will become automatic. Practice your key words by calling them off as you would the alphabet; ape, bee, sea, December, etc. You can learn your entire list in half an hour.

Here is an easy way to quickly recall the number of your key word without memorizing each one. Just remember the meaningless word EJOTY. E is the fifth letter of the alphabet, J is the tenth letter, O is the fifteenth, T is the twentieth, and Y is the twenty-fifth. If you wish to know the seventeenth letter in the alphabet, just think of EJOTY. Since O is fifteenth, reciting the alphabet for two more letters brings you to Q, the seventeenth letter.

Of course, do all your associating quietly. Never let your audience know that you came prepared with a memorized list. Let them think that you are a genius! However, your classmates might find the scientific explanation extremely fascinating and thank you for it.

Suggested List

A. ape
B. bee
C. sea
D. December
E. eel
F. fairy
G. glacier
H. heaven
I. eye
J. jay bird
K. key
L. elevator
M. mother
N. envelope
O. oak tree
P. peace pipe
Q. queen
R. railroad crossing
S. steam iron
T. tea party
U. U-boat
V. volcano
W. washtub
X. X-ray
Y. YMCA

Where is your skin most "touchy"?

Have you ever seen a blind person read a Braille (brayl) book by running his sensitive finger tips along raised dots on the pages? Strangely enough, he probably would not be able to read the book with the tip of his elbow or his ear lobe.

The reason is that the nerve cells specially designed for touch are not distributed equally in the skin all over the body. Places which are used for touching, such as fingers, have the touch spots closer together. Lips, for example, have these sensitive spots ten times closer than the middle of the back.

You can find out where you have more nerve endings sensitive to touch by making a simple tester with a hairpin. First, open the ends so that they are ¼ inch apart. Touch your finger tips with these ends. If you feel two points of contact, then bring the ends together. Continue making the distance smaller until you obtain the closest distance that your finger tips can still feel two points. Your tester is now ready to use.

You can do the experiment yourself, but it will have more value if you get a friend to help you. Close your eyes and ask him to touch various parts of your body. You must tell him whether you feel one or two points. Instruct him to try to fool you by touching you with only one end from time to time. But of course, every tested spot must finally be touched by both ends. Make a list of the areas of the body which are tested and indicate whether you feel one or two points.

You can make a more permanent tester which does not change its adjustment as easily as a hairpin. Simply stick two blunt pins through a ½-inch slice of cork.

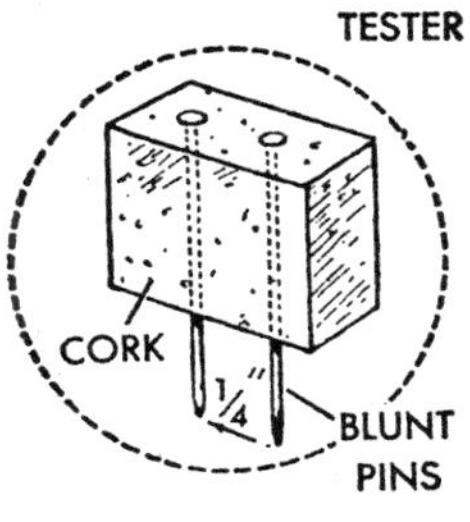

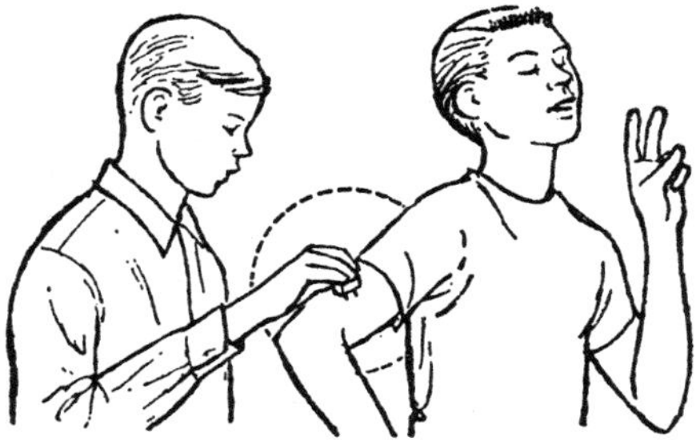

When your finger tips get cold, do they lose their ability to distinguish the two points of your tester? Keep your hand in ice water until it feels chilled and then test it.

Do you think that your tongue has many touch spots

close together? Use a washed tester and check yourself.

Are your toes as sensitive as your fingers?

The skin has an outer covering called the EPIDERMIS eh-pi-DER-miss). But it is in the inner layer of the skin, called the DERMIS, that the sense organs of touch are found. When a nerve ending sensitive to touch is stimulated, an electrical signal goes to a special part of the brain called the touch center. When you feel two points of the tester, it is probably touching at least two nerve endings. When you feel one point you may be touching only one nerve ending.

In addition to the sense organs for touch, there are also other special sense organs for pressure, pain, and separate nerve endings for registering heat and cold. Nerves may be at different depths. That is why a light touch stimulates the nerve endings for touch while a heavier touch activates the nerve endings for pressure which lie deeper in the skin.

Do the following experiment to convince yourself that the heat and cold nerve cells are two different sense organs.

Mark off an area about 1 inch square on the back of your hand with a pencil. Use a sharpened pencil point to touch each spot of skin in this area with the same pressure. Do this in regular rows and columns. Each contact should be about $\frac{1}{16}$ inch apart. Count the number of cool spots and hot spots resulting from the touch of the pencil. These are surprisingly easy to distinguish. Do this for different parts of the body.

YOUR BODY—*More to find out:*

Can you see your pulse move?

Your pulse is a movement of your blood vessels when a wave of blood is pumped from the heart. It occurs in an artery since this blood vessel carries blood *from* the heart. One familiar place where an artery comes close to the skin is at the wrist. You can feel the throbbing by using the finger tips of the other hand.

In order to *see* this tiny movement, you must invent some way of making this motion more apparent. One way is to place upon the pulse a very small piece from a broken mirror, about ¼ inch wide. Allow sunlight or a flashlight beam to be reflected from the mirror to a wall or ceiling. You should get a noticeable movement.

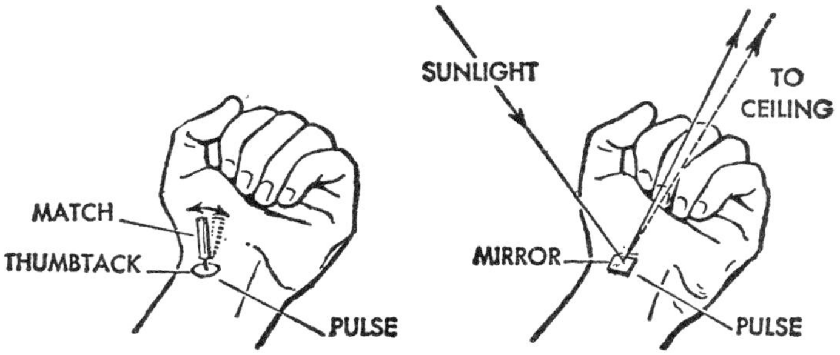

Another method is to stick the end of a wooden match on the point of a thumb tack. Stand the head of the tack upon the pulse. The match will bob in time with your pulse. Can you think of other ways of making this small motion more recognizable?

How fast do fingernails grow?

What kind of a reference mark can you make at the base of the nail which will not wash off or wear off? It will have to last for many months or be renewed as the nail grows upward. Is there some paint or chemical which causes a stubborn stain? Use the edge of a small file or a coping saw to etch a light line you can watch.

Do all fingernails and toenails grow at the same rate? Take measurements regularly.

Are any of your friends color blind?

Do any of your friends have trouble selecting the proper colors of ties or socks? Do some fail to distinguish a red from a green traffic light? About 7 per cent of all males and ½ per cent of females have some defect in color vision. It is inherited.

You can test your classmates, friends, and relatives by asking them to name the colors of objects in your test kit. This consists of differently colored pieces of wool, cloth, paper, and other objects. Some biology and other books may have special charts for testing color blindness. Ask your doctor or optometrist where to obtain these charts.

Calculate the percentages of males and females who are color blind. Find out whether anybody else in the family has this defect among those you test. Chart your results.

Are both your feet the same size?

Each foot contains twenty-six bones connected by ligaments so that they form a springy arch. The bones and flesh have millions of growing cells. The foot size is determined by the height of the arch, the space between the bones, the rate of cell growth and many other variable conditions. Would it not be a miracle if both feet were *exactly* the same size?

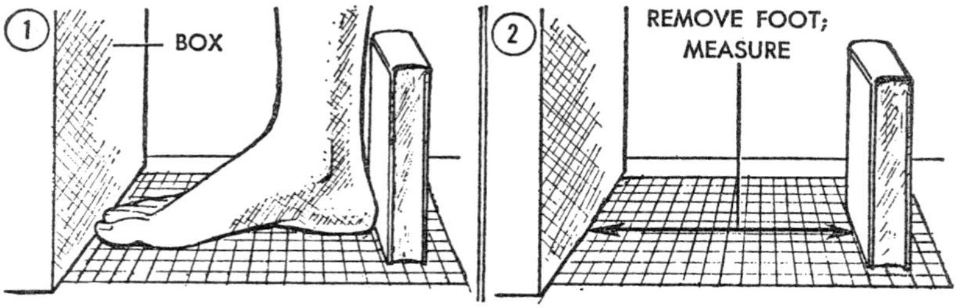

Devise an accurate way to measure the sizes of both feet of your friends. One way is to place the foot flat on the floor with the toe against a wall which has no base molding. The side of a large box can be used instead of the wall. Press a vertically-held book or another box against the heel. Remove only the foot and measure the distance. A large sheet of cardboard ruled off in eighths of inches will be found helpful. All measurements should be made with the people *standing*.

You will discover some great differences between the left and right feet. Keep records of subjects' names and sizes of each foot. Shoe salesmen can give you much information about this human problem.

What is the best way to take fingerprints?

It is interesting to study fingerprints. But since most of you do not own a professional ink pad, the hobby can get mighty messy! Also, with a homemade ink pad, the print is rarely a perfectly clear one. Can you invent a reliable, clean, fingerprinting device?

There are many possibilities to explore: different colored inks and other household fluids; fine dusts such as scraped charcoal, graphite from pencils, talcum powder, corn starch; oily substances. Experiment with different surfaces such as glass, glazed paper, smooth, but absorbent paper, mirrors.

Here is an idea which has to be worked out. Gently rub a finger on a pencil-smudged area. Then touch the sticky part of cellophane tape. You should get a beautiful print. However, the correct position to read the fingerprint is with the sticky side facing up. What can be done to cover the sticky side?

How much mineral matter is in bone?

Collect many chicken bones. Clean off the gristle. Heat in a large tin can for an hour or so, using the largest flame. Cool, and now you can easily chop up the bones with a hammer. Reheat. Most of what remains are compounds of calcium, phosphorus, magnesium, and other minerals. If you weigh the bones before and after burning, you can get an idea of the percentage of minerals in bones.

LIVING THINGS

Can you make a frog hibernate?

Around October, when the temperature starts to fall, frogs move about more and more slowly. Finally, each one becomes extremely sluggish and "goes to sleep" for the winter in a sheltered place. Biologists call this process HIBERNATION (hy-ber-NAY-shon). Most frogs hibernate in the ground. Some crawl into the soft mud or under a stone at the bottom of a brook or pond.

During the cold days ahead, the frog is barely alive. All its body's functions maintain the minimum amount of energy necessary to prevent death from starvation. The heart works very slowly, while the lungs are not used at all. Instead, the moist skin does all the breathing. The frog lives on its stored up fat. The temperature of this cold-blooded animal is now only a degree or so above its surroundings.

Hibernation saves the lives of these animals which would otherwise freeze or become so slow that they would be unable to escape their enemies. It is an interesting coincidence too, that their insect food supply becomes scarce at the same time.

Your exciting research problem is to find out at what temperature a frog starts "digging in" or behaving as a hibernating frog would at the bottom of a pond. Follow the simple technique which is given below, and you will be able to make a frog "hibernate" any time you want it to.

Place a live frog into a jar almost completely filled with water and containing a thermometer. A good species to use is a leopard or a green frog. If you live near a pond you can catch one. If not, borrow one from a school. Place the jar in a pan containing ice cubes. Keep watching the thermometer and the frog.

If the water is at room temperature when you start, the frog will probably go down to the bottom of the jar,

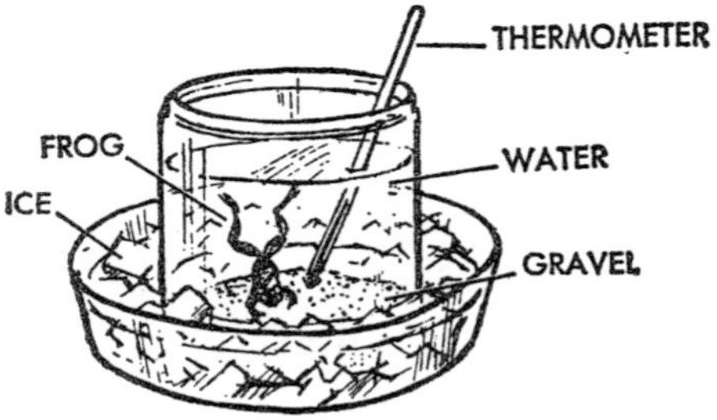

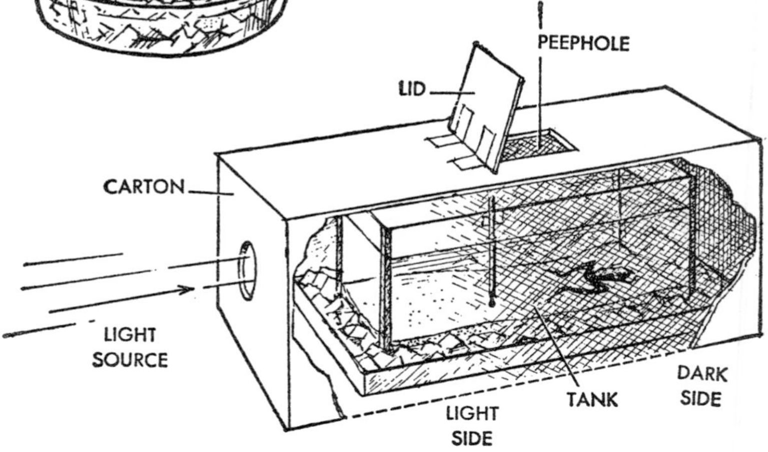

and then up again. As the temperature of the water keeps falling, the frog becomes more and more sluggish and stays near the bottom. You can also watch how the breathing becomes slower as the water cools, by observing the pulselike throat movements of the frog. Near 46 degrees Fahrenheit, the frog may try to hide by nosing its way under anything within reach. At about 41 degrees all motion usually ceases. Whether moving or quiet, a hibernating frog will hold its head pointed downward, and the hind legs are sprawled outward.

Record the temperature at which hibernation occurs. Do it several times. Is the temperature always the same? Record the temperature at which the frog returns to normal. Remove the jar from the pan of ice when hibernation occurs. Do not freeze the water. If you do this experiment correctly, the frog is not hurt.

Here are several variations. Try small jars, large jars and also an aquarium tank. Place different flat objects on the bottom. See what happens when the bottom has a 2-inch layer of gravel. You might also try these experiments with newts.

Another interesting variation is to arrange to have one side of a long shallow tank dark, while the other side is light. Frogs generally like dark places to hibernate, and when placed in water about 46 degrees Fahrenheit, will go toward the dark side. In warmer water you will find that the frog usually prefers the light side. Try admitting light through a big hole on one side and through a small hole on the other side.

Are only the north sides of trees green?

Books on camping, scouting, and even biological subjects often make the statement that when there is a greenish growth or stain on the trunk of a tree, it is on the north side. Indeed, some even suggest that a lost person can get his bearings by knowing this information. It can be extremely interesting to a young scientist to check the accuracy of these statements in his locality.

When the greenish growth is moss, you can no doubt recognize it. Sometimes however, you see a greenish stain on the bark. It is probably caused by an ALGA (AL-ga) called PLEUROCOCCUS (PLOORO-cock-us). Very frequently you see a scaly or leaflike crust on the bark. This consists of LICHENS (LYE-kenz). A lichen is really not one plant, but a combination of two, a FUNGUS (FUN-gus) and an alga.

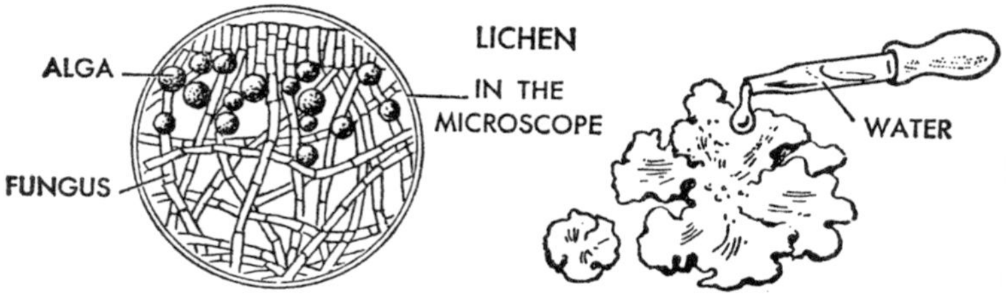

The fungus consists of very tiny strands of colorless cells. It has no chlorophyll and therefore cannot make its own food. But, living in this matting of threads is the alga which is green, and therefore can manufacture sugar and starch. The fungus depends upon this food.

In this mutual benefit partnership, the fungus does its share by absorbing and holding water, supplying minerals and also anchoring itself. You can see, by wetting a clump of lichens, that it absorbs water like a sponge. Use a medicine dropper filled with water. Obtain different lichens and examine them under a microscope. Lichens are extremely hardy. They are even found in the polar regions, often supplying reindeer with food when hardly anything else is available. It is no wonder that lichens can grow on the north sides of trees and rocks.

Since moisture is a critical necessity for moss and lichens, they thrive best where they can hold the moisture longest. The north and northeast sides of trees have most shade. Therefore, there is less evaporation and these hardy plants are very frequently found there.

However, more important than retaining moisture is receiving new moisture. Most investigators believe that moss and lichens will grow best on the side receiving moisture for the longest time.

Make a study of your locality. From which direction do rain-bearing winds come? Also, what direction is shaded for the longest time? Combine these two facts and see whether the lichens are growing where you think they should.

Take walks to different locations. Get accurate directions by using a compass. Keep good records. See if all trees in a certain region have lichens or moss similarly located on the tree trunks or rocks. Are there places where the rain-bearing winds are reflected, and come from a different direction? Does this have any effect upon the nearby trees?

Do you find any differences in lichen growth caused by obstructing tall buildings, winds over lakes, land and sea winds at the seashore, streets running in certain directions, rain spray from a nearby wall?

Can you explain why the trunks of some trees are green on all sides? Can you find a reason why some trees show no lichens?

From your studies, do you think that a lost hiker can always find north by observing lichens on tree trunks?

How do dandelions ruin lawns?

A dandelion is one of our most beautiful flowers. It gets its name from the French, *dent de lion,* which means, "tooth of the lion." But most people see very little resemblance of its petals or lancelike leaves to a lion's tooth.

It is one of the earliest flowers of the season and of course, all the children love to pick them. Many gardeners even grow them in pots, just to show off their golden color. The Apache Indians used to scour the countryside for them, not only for their beauty, but because they were considered delicious when eaten. And even today many people eat dandelions.

Despite all the virtues of this interesting plant, homeowners hate to see its flowers in the lawns because they spoil the grass. It is also very difficult to get rid of them.

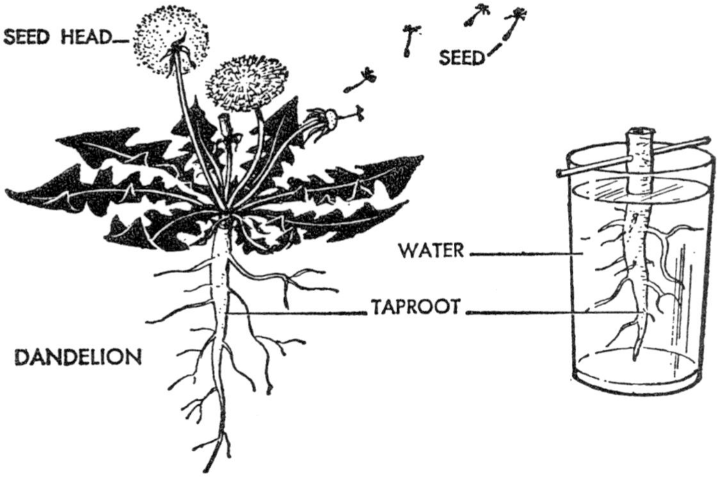

They are called weeds, since that is what plants are when they grow in unwanted places at the expense of more desirable plants.

You can study the dandelion and learn several reasons why it is so hardy and also undesirable in a lawn. Do the following:

1. Find a large dandelion plant in a lawn. See how the leaves are arranged in a circle close to the ground. This kind of roselike shape allows each leaf to get the maximum amount of sunlight. But, at the same time it prevents the sun from reaching the grass underneath, and the grass dies off.

Can you work out the following simple problem? How many square inches of grass does a large dandelion plant kill? (HINT: The area of a circle is equal to πr^2.)

Just think of all the grass which is ruined by a dozen dandelions on a small lawn.

Dig up a plant and see why the rosette is formed by the leaves. See how the larger leaves are on the outside and the shorter ones are near the center. The leaves on the outside also have longer stems.

When a lawn mower goes over the plant the blades pass right over most of the leaves. This is why home-owners have to do extra work to remove the plant.

2. To learn this first hand, try to pull up a dandelion plant with your fingers. It is almost impossible to do so. That is because there is a large main root, called a taproot, which goes down fairly deep. This anchors the plant very securely. Use a trowel and dig around the plant until the entire root is extracted. Measure its length. Do this with several large dandelions.

A homeowner must remove the entire taproot. Just cutting the leaves and short stems off the taproot, or cutting part of the taproot, will only cause a future growth more troublesome than before.

To demonstrate this, dig up several plants. Place the taproots with the smaller roots still on them, about half-way into a glass of water. Keep the pointed sides of the roots down. Remove all the leaves. (See illustration.) You will probably find new leaves growing from the top of the root in several weeks.

Another reason for removing taproots from the ground is that a dandelion is a perennial plant. Its root does not die in the winter and will reblossom in spring.

3. Count the number of yellow petals. Each one is a part of a separate flower which will produce a seed. Dandelions, like daisies, consist of many flowers very closely packed together in what is known as a FLOWER HEAD. Use a magnifying lens to examine this. You can also try to find out whether all dandelions have the same number of petals. If not, how wide is the variation?

4. When the flower withers, and the familiar SEED HEAD appears, you can see many dozens of seeds. Each seed is attached to a parachute made of silky fibers. The wind blows the seeds everywhere.

Count the number of seeds in a seed head. Remember that all these came from only one flower head. Blow on it. See the seeds sail away. Try to grow some seeds in a pot. You might place a tag on a dandelion flower head whose individual petals (flowers) have been counted. When it forms a seed head, count the seeds. Did every flower produce a seed?

5. Stake off two areas on a lawn each 4 feet square. Do this on some remote section in back of the house. Remove all weeds from one area whenever they appear. Allow the other area to grow wild. However, you must water each area equally, and mow each at the same time. All other conditions must also be the same for both.

Do you notice any difference in the appearance of each area? What conclusions can you draw about weeds? Find out how special weed killers such as 2–4D are applied to lawns.

How can a spider's web be collected?

J. Henri Fabre, the great French naturalist used to call himself "a self-appointed inspector of spiders' webs." He would spend many enjoyable days watching every detail in their fantastic construction. On your walks, you too can discover first hand the amazing engineering and architectural skill that goes into a web.

When you find a finished web you can photograph it, if the light is right. Another way to record it is to make a careful drawing in your notebook. Be sure that you have the correct numbers of spokes, spiral threads, and supporting strands and that all the angles are represented accurately.

It is also possible to bring back the web itself so that you can display and study it. Prepare yourself with a large sheet of stiff cardboard, preferably black, which is used for posters. Also bring along a small bottle of shellac, glue, or fast drying mucilage. The cement and the lacquer (dope) used in model airplane work are also good for this purpose.

When you find a suitable web, hold the cardboard up near the web for measurement. Cut the cardboard slightly smaller than the web. Now paint the edges, or, if you wish, a 1-inch margin on the cardboard, with the shellac or other adhesive. Press this prepared cardboard carefully against the web. Hold it there until you are sure that the threads of the web have stuck to the cardboard. Then cut the web around the cardboard. You

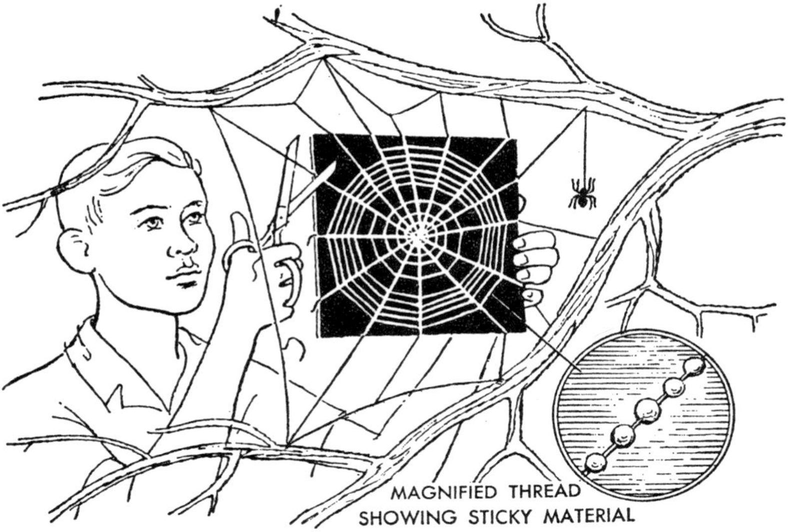

should now have the web suspended on the cardboard exactly as it hung in its natural condition. Perhaps you can think of some way to put a transparent dust cover over it, such as cellophane or glass.

Not all spiders spin webs, but those that do, use the sticky strands for trapping insects. The spider hangs head down from the center, or waits at the edge of the web until an insect lands in the web. In a twinkling the spider is upon it, enmeshing it in more strands of silk and biting it to death, and sucking its blood.

The silk used for the web comes from three pairs of spinnerets on the tip of the abdomen. Each spinneret consists of hundreds of microscopic tubes through which the liquid silk flows from silk glands. It hardens immediately in the air. The hind legs have a comblike arrangement with which the spider pulls and guides the silk. Spider web silk is only about four-thousandths of an inch in diameter and is extremely strong for its size. It is used as cross hairs in optical instruments such as range finders and surveyors' telescopes.

You may be fortunate enough to observe a spider in the process of building a web. Here is how one species does it. First it stretches a few threads across an open area. These act as a framework or foundation. Then a special thread containing a center white dot is stretched across. Through this, a few spokes are laid in one direction and then, to avoid strain upon the incompleted web, several spokes are laid in the opposite direction. When all the spokes are in place, the spider starts, from the center, to put in the spiral threads between the spokes. When the spirals are done and the spider is on the edge of the web, it retraces its path back to the center.

As it returns, there is left on the spiral threads the extremely sticky material which traps the prey. The spokes are left dry. The spider does not get stuck in its own snare because it knows where to walk!

Spiders are called ARACHNIDS (uh-RACK-nids) and are not insects. They have eight legs while insects have only six. Almost all spiders are harmless. The female black widow spider found mainly in the southern states is the only spider which can inflict a very dangerous bite. She has a red "hour-glass" mark on the bottom side.

Read about spiders in the library. Learn more on your walks by doing the following:

Observe the beads of dew on webs in the early morning. Throw a fly into a web. Where does the spider come from? What is done with the fly? Use a magnifying lens to examine the sticky globules on the spiral strands. Keep records of position of webs and direction of prevailing winds. Spiders usually do not build webs against the wind. Watch a spider go up and down on a silky thread when it works or wishes to "balloon" away. This is called a dragline.

Touch various parts of a web with a twig. Which are sticky and which are dry? Preserve spiders in 70 per cent (rubbing) alcohol. Cover vial tightly. Look for a spider with a small sac-shaped case. It is the egg case. Keep it in a warm place. When young hatch, they look like parents and eat each other.

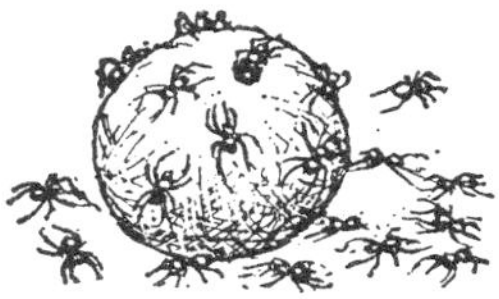

Notice that spiral threads are not curved. Spiders can only make straight lines and angles. Are there any gaps in the web through which an insect can fly? Place an upturned jar on a basin of water. Place a spider on it. It cannot escape except by being wafted away on a silky parachute it makes. See if it does this.

How can plants be watered during your absence?

A constant problem at school or at home is how to water plants during weekends or vacations. Many people merely place the flower pots into a basin containing several inches of water. However, most plants rarely thrive when they remain in water for long periods.

Experiment with the following method using plastic bags which do not have holes. Place a bag over a thoroughly watered plant. Use sticks, wire hangers or other means to keep the sides of the bag from touching the leaves. Bend the opening of the bag under the pot to prevent the escape of moisture. If you find that some water does get lost, stand a tumbler of water inside the bag. Keep this set-up in the sun for many weeks and study the effects, if any, upon the plant.

Another method, which should give you many days of careful research, is based upon CAPILLARY ACTION (CAP-il-lar-ee). This is a process where water creeps along or rises in porous materials or narrow tubes.

Place the plant to be watered some place where it can get good drainage. Place a large pail or jar of water next to it, but *above* it. Set the end of a long narrow cloth in the water, down to the bottom of the pail. The cloth should extend over the edge of the pail and down to a short distance over the flower pot.

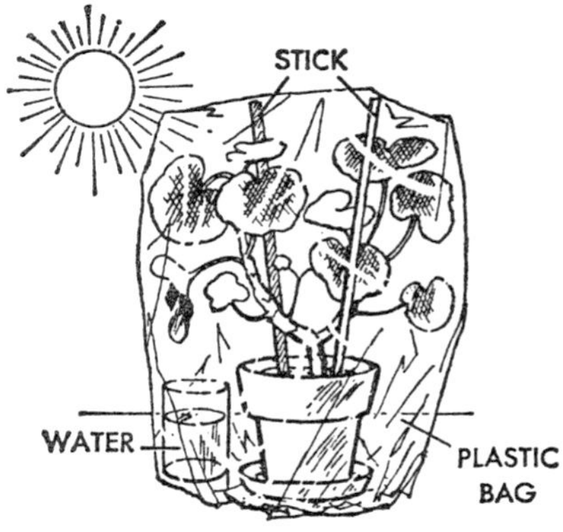

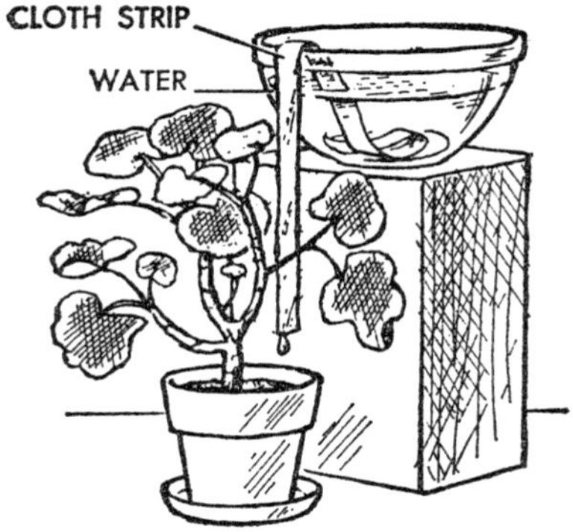

After a while the cloth will get wet, and water will drip into the pot. Experiment with different kinds of fabrics. By regulating the width of the cloth, the proper amount of water will fall on the plant for the longest time.

Can you invent other ways of keeping plants alive over extended holidays?

Can the shape of a tree indicate direction?

You have often seen pictures of a wind-swept tree. If you knew the direction of the prevailing winds you could tell the compass directions just by looking at this shape.

The main trunk of a tree is usually not exactly in the center of all the growth around it. (See illus., p. 120.) The side facing the wind may have smaller and fewer branches. This is often quite obvious in the wintertime.

On your walks, see whether trees in your locality show this greater growth on the wind-sheltered side. Do not base your conclusions only upon a few trees. Always choose trees which are out in the open and not sheltered by hills, houses, or other trees.

The American Indians believed that the very tops of evergreen trees pointed toward the direction from which the sun shone for the longest periods; that is, in a southerly direction. How much truth is there to this bit of Indian lore?

Another interesting study is to examine the unremoved stumps of cut-down trees. The annual rings are rarely perfect circles with the heart in the center of the stump. Instead, the center of the distorted rings is more toward the south and east side. The rings too, are farther apart toward the north and closer together toward the south.

The Indian scouts believed that tree trunks were thickest on the north side. Nobody knows the explanation for this. Can it be that a tree is thickest on the side which has to be stronger because of winds?

Keep records of the observations you make. Write the location of the tree stump, the appearance of the distorted rings and also the compass directions. Then draw your own conclusions.

How many living things can you find in soil?

It is amazing how many living things there are in the soil besides earthworms. Most of them can only be seen with a microscope, but you can chalk up a high score using only your naked eyes and perhaps a small magnifying glass.

For this outdoor study take along a small spade, some large heavy shopping bags, several large sheets of white paper or newspaper, a ruler, and some plastic vials with covers. Of course, do not forget your pencil, notebook, and magnifying glass.

Collect the earth from several locations so that you can compare your findings. Get one sample of earth from below the leaves lying in the woods, another from a grassy field, and also from a badly eroded field. Measure off an area 1 foot square. Dig out the soil to a depth of 3 inches and place it into bags. Then spread out the soil on the white paper or newspaper and start your count. Separate the plants and animals to the best of your ability. A section of window screening might be found helpful in counting smaller animals. Used as a sieve it allows the earth to pass through while holding back the animals. Place anything interesting in a vial for future study.

Which soil has most living things? Do living things in the soil make the soil looser? Estimate the total number of animals in an acre of soil. (There are 43,560 square feet in one acre.)

How long can an aquarium be "balanced"?

A balanced aquarium is one whose cover can be permanently sealed to the air. All the living things in the water will continue to live because the plants produce in exact quantity what the animals need, while the animals produce just what the plants need. In other words, no food is ever added from the outside.

A green plant manufactures starch in sunlight by a process called PHOTOSYNTHESIS (foto-SIN-thesis). It takes in carbon dioxide given off by the water animals. At the same time the plant gives off oxygen which is used by the animals. The animals also eat parts of the plant and they drop waste products, which become fertilizer for the plant.

Many biologists believe it is impossible to keep a balanced aquarium indefinitely. Sooner or later something multiplies too rapidly or something dies. This leads to pollution of the water and the deaths of plants or animals, unless the water is changed.

Your research problem is to make various miniature balanced aquaria and find the best combination of conditions. You can use test tubes sealed with corks or small glass jars which can be sealed absolutely airtight. Put in small fish, small snails, or both. Use a good

oxygen-producing water plant, such as elodea. Place some fine sand in the bottom of some containers.

A simple miniature balanced aquarium can be attempted by placing a small snail and a sprig of elodea in a test tube of water. Leave an air space of about 1 inch. Cork tightly and seal with wax. Place it in moderate sunlight. Label it and keep good records of the clarity of the water, amount of plant growth, the apparent health of the snail.

Start a string of other aquaria, each one with a slight variation in the amounts of plant and animal life, sunlight, gravel, or anything which you think might have an effect.

Can you keep a balanced aquarium for six months?

How does an earthworm react to electricity?

Dig for earthworms in moist, rich soil or find them on the surface after a warm rain, especially in the evening.

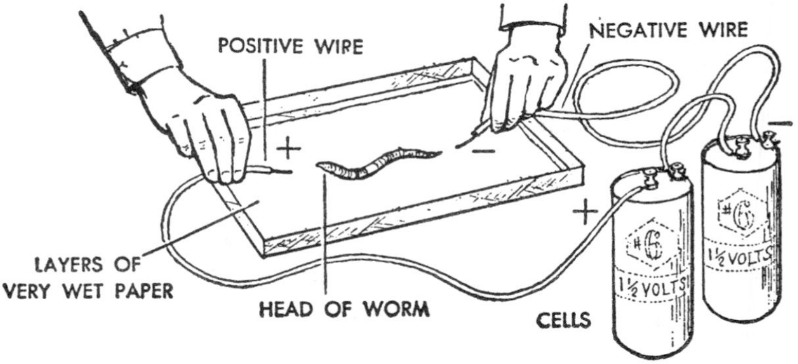

Place a worm on several layers of *very wet* paper. Hook up two dry cells in series by connecting the positive post of one cell to the negative post of the other. To the other two posts attach insulated wires about 15 inches long and with the insulation removed from the ends.

Hold the end of the wire which is connected to the positive post. Touch it to the wet paper, about an inch from the worm's head. Also touch the tip of the negative wire to the paper, about 1 inch from the tail end of the worm. The worm will probably draw itself together like an accordion.

Reverse the position of the wire and the worm should slowly become elongated. Will every worm you have contract when the positive wire is near the head?

Again place the wires on the paper, but this time one on each side of the worm. Most likely the worm will turn its head *toward* the wire from the negative post. Scientists do not know the exact reason why a worm's

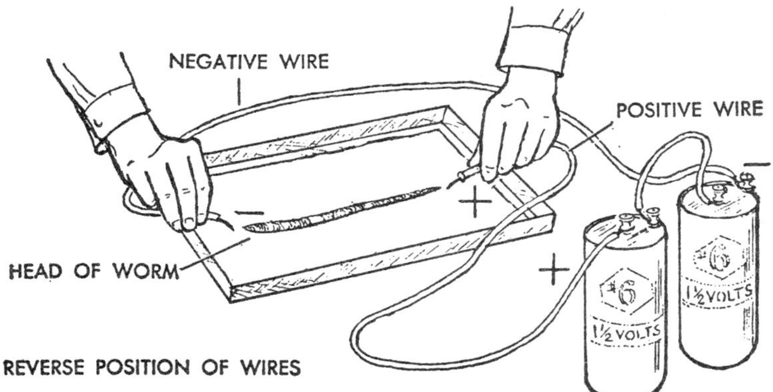

nervous system should react to electricity with such polarity.

What are the effects of higher or lower voltages? What happens when the paper on which the worm rests is moistened with a very weak salt solution?

You can see how a worm's nervous system also reacts to chemicals. Remove the worm from the wet paper. Moisten a piece of paper with vinegar or household ammonia. Bring it near the head of the worm, but do not touch it. The head should move away quickly. Test whether any other parts of the body also move away.

Worms also react to strong light. Keep a worm in a dark place for a while. Then shine a flashlight on it. It will move away, even though a worm has no eyes. Find out a worm's responses to noise, heat, cold, winds, and other annoyances.

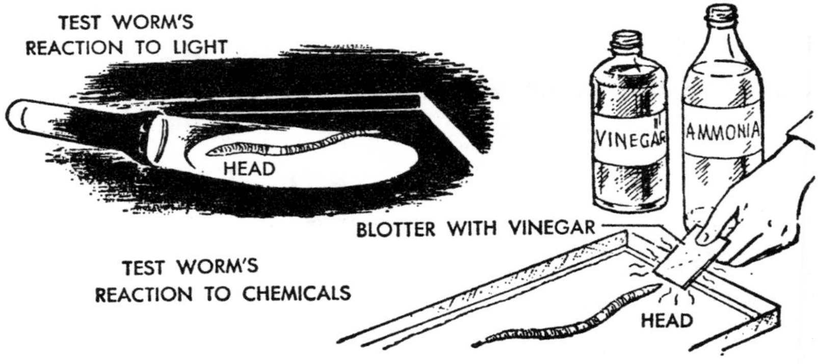

SOUND AND LIGHT

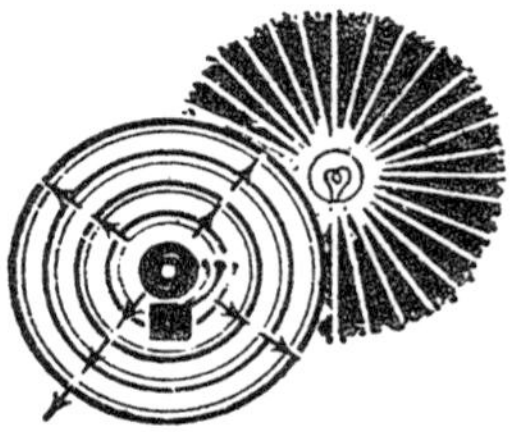

How can we see sound vibrations?

Perhaps you have seen an electronics engineer or a technician working with an OSCILLOSCOPE (ahs-SILL-a-scope). This is a device which creates different kinds of curves, waves, loops, and lines on a small screen. These patterns vary according to the type of electrical signals fed into the amplifier.

However, the "oscilloscope" which you are going to make is not going to be electrical. Instead, it will produce various curves and shapes when different sound vibrations cause a spot of light to dance on the wall.

Obtain a small tin can which has already been opened. Use a can opener to remove the other end without leaving any jagged edges. Stretch the bottom half of a toy balloon over the end of the can as on a drum. Use string or rubber bands to hold the rubber in place.

From a small pocket mirror cut a piece about ¼ inch square. If you cannot get someone to cut it for you, then carefully break the mirror in a paper bag. Look for a small suitable piece. Glue this on the rubber membrane with the mirror side up, and about one-third of the diameter from one edge.

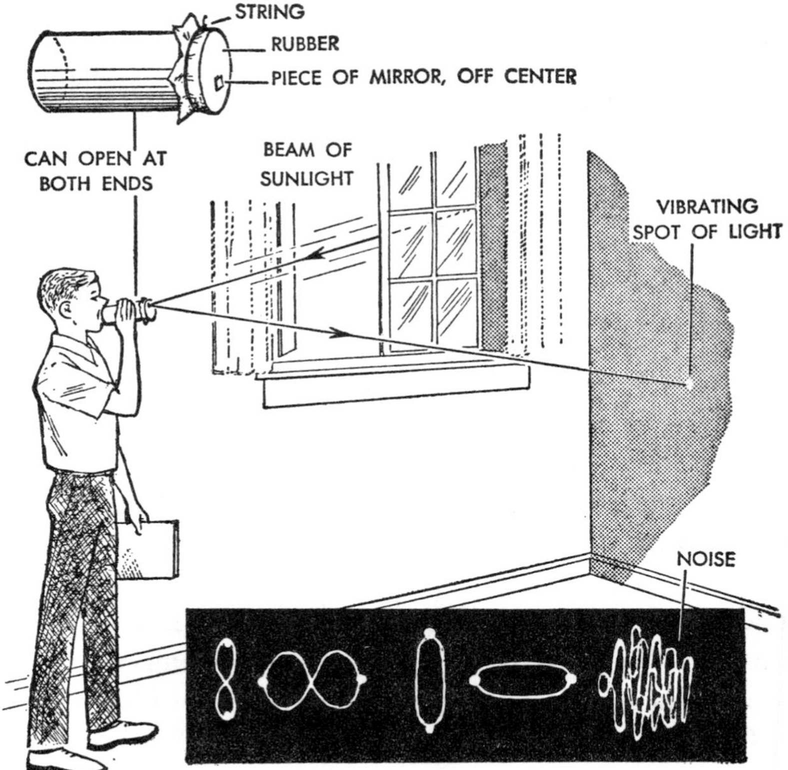

SOME FIGURES SEEN ON WALL

Place yourself in such a position that a beam of strong sunlight will be reflected as a spot of light, from the tiny mirror to the wall. Or, if you wish, you may hold a small white cardboard in your hand to catch the reflected spot.

Put your mouth close to the open end of the can and make singing sounds. Notice the different patterns which are formed. Among the shapes will be ovals, figure-eight loops, lines, and also combinations of curves and lines.

The vibration of your vocal cords causes the air in the can to vibrate. This makes the rubber and the mirror vibrate. The sunbeam reflecting from the vibrating mirror produces the patterns you see on the wall.

Make continuous singing sounds—ee, eh, ah, oh, ooh, ay, aw, eye, etc. See how each sound has its own characteristic shape. When a sound is sung loud, the shape increases in size. Keep notes of your findings.

It is interesting to observe that the thinner and more definite patterns are produced by musical notes. This is because musical sounds consist of even vibrations, a definite number per second. When noisy sounds are uttered, the pattern of the vibrations is not a simple one, because a noise is a mixture of many irregular vibrations. A sound which is not sung, but almost talked, does not have a simple pattern either, because of the many different vibrations.

Sing each sound up the scale and observe that the pattern may rotate a little more with each rise in pitch. Humming sounds do not produce good shapes because the vocal cords are not involved as much as in other sounds.

Here are more suggestions for experimentation with this sound apparatus. While you are making the sounds, see what happens to the shapes when you rotate the can. Try varying the size of the can, the tightness of the rubber and the location of the mirror on the membrane.

What happens when the mirror is placed exactly in the center? What shapes will you get if you glue two mirrors on the rubber in different places?

You might try using a tiny circular mirror often found on ornamental work. A glazier will be glad to give you several. You can also find them on cheap costume jewelry.

This nonelectric "oscilloscope" can be used at night by shining a flashlight beam on the mirror. Notice that the farther away you are from the screen, the larger becomes the pattern. However, the reflection also gets dimmer.

How well can you judge direction of sounds?

You probably already know that you can judge the direction and distance of an object because each of your eyes sees from a slightly different angle. This is called STEREOSCOPIC (steer-ee-oh-SCOP-ic) VISION. Similarly, because of the location of your two ears you can identify the direction of a sound. This kind of hearing is known as BINAURAL (by-NAW-rel), meaning "two ear."

When you hear a sound coming from the right side, then the right ear hears the sound a fraction of a second before the left ear. Since sound travels in waves, one ear hears a different part of the same wave. Your ears then change the sound waves into nerve impulses traveling to the hearing center of the brain. Your brain interprets the two slightly different, and out-of-step impulses as a sound originating to your right.

You judge the direction of a sound best by unconsciously moving your head around, until both your ears hear as differently as they can. The brain is confused by sounds originating in back of us, or above us.

Do the following experiment to show how difficult it is to know the direction of a sound when only one ear is used.

Blindfold a friend and have him sit in a chair in the center of a room. Have him block one ear with his hand, or carefully put absorbent cotton into the ear. Tell him that you are going to tap a pencil against a bottle in different parts of the room. Ask him to point to the place where he thinks the sound is coming from. Move around quietly. How accurately can one determine direction with only one ear?

After a while do the same experiment while the subject has both ears open. Make the tinkling sound in front of your friend; also in back, left, right, above, below. Which directions are hardest for him to be sure of?

It took years of experience for your brain to develop its keenness in judging directions of sounds. You can demonstrate the effect of undoing all the previous learning of your brain by means of the following experiment:

Obtain two 15-inch lengths of rubber tubing, and two funnels. Place a funnel into one end of each tube. Insert the free end of one tube into your right ear. Now bend the tube across your face so that the funnel is facing the left side. Put the end of the other tube into the left ear and bend it so sounds will be received from the right. Use string or adhesive tape to hold together the tubes crossing your face.

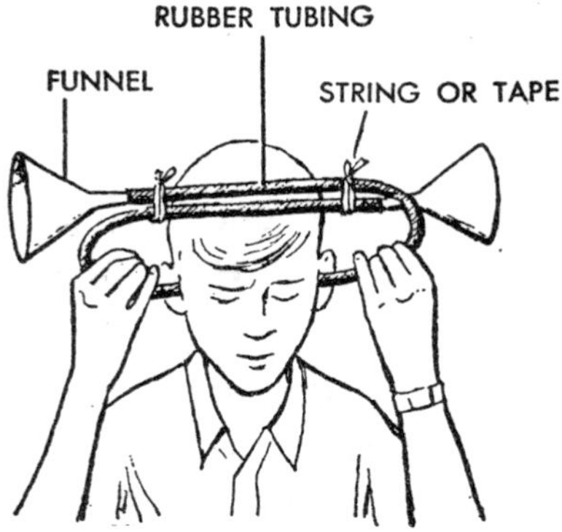

Use this device to determine directions of sounds. It is quite frustrating to do this accurately with your eyes open. But with your eyes closed, it becomes utterly confusing and alarming.

This shows that the ears and brain can be fooled. A ventriloquist takes advantage of this fact by making his voice appear to come from his dummy. Sight and your imagination do the rest.

Can you identify types of eyeglasses?

Would you like to make a very big impression upon a friend or relative who wears eyeglasses? Simply ask the person to remove the glasses. Hold these at arm's length and look through them while moving them to the left and to the right. Now, with a wise look upon your face you can announce whether the wearer is nearsighted or farsighted.

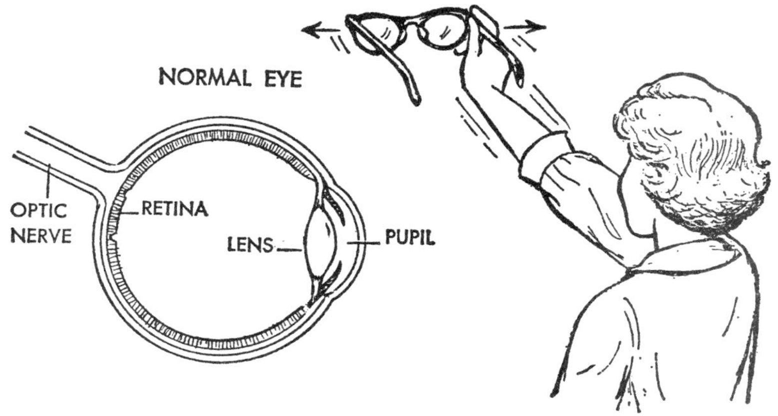

Before you can do this interesting stunt there are several facts you must know about eyes and lenses.

Since the eye is like a camera, it contains a lens system. This is capable of forming a clear image upon a sheet of nerve endings called the RETINA (RET-in-a). The optic nerve then sends messages to the brain where these impulses are interpreted as sight.

A normal eye has a certain shape enabling a clear image to be formed exactly upon the retina. However, an eye which is nearsighted usually has a long shape and an image comes to a sharp focus *before* the retina. Therefore a nearsighted person has blurred vision.

NEAR-SIGHTED EYE

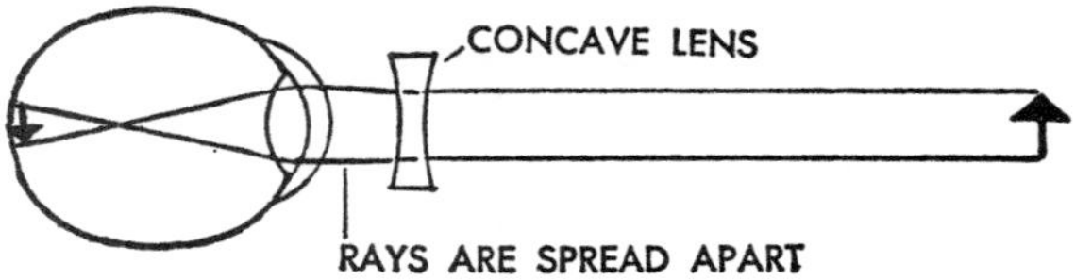

Another imperfect eye is the farsighted one. Here, because the eyeball is too short, the sharp focus occurs *in back of the* retina. Again, vision is blurred.

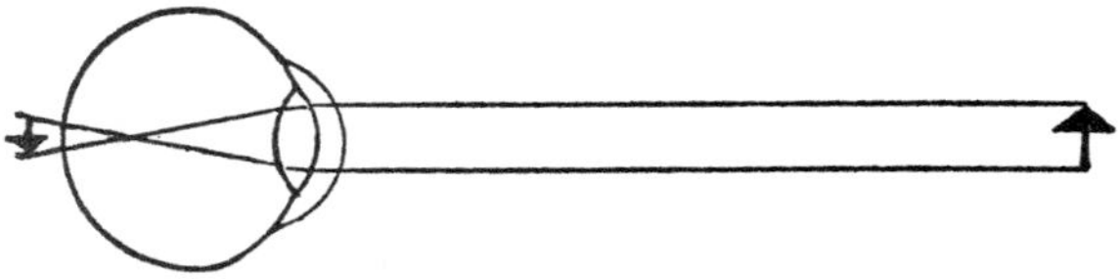

FAR-SIGHTED EYE

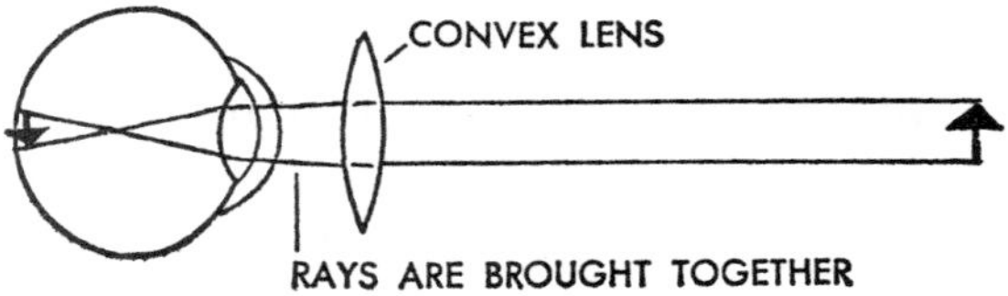

An optometrist (op-TOM-eh-trist) can place certain lenses in front of the eye to make the focus occur exactly upon the retina.

One of these lenses is a CONVEX lens. This is thicker at the center than at the edges. It makes rays of light come together. When held in front of a farsighted eye it bends rays slightly together before they enter the eye, as you can see in the diagram. The effect in the eye is to shorten the focusing distance. The sharp focus now occurs inside the eye and upon the retina.

If you have ever played with a convex lens such as a magnifying lens, you must have seen how it inverts an image which is cast upon a wall. Everything is upside down and left to right. Also, anything you look at through a convex lens will move in a direction which is opposite to the direction that the lens moves.

Another lens used in eyeglasses is a CONCAVE lens. This is thinner at the center than at the edges. A concave lens is used to correct a nearsighted eye because it spreads the light rays before they enter the eye. As you can see by again studying the diagram, the effect is to increase the focal distance so that the sharp image now falls upon the retina.

A concave lens does not invert images. Moving the lens causes everything you see through it to move in the same direction as the lens.

To review:

1. Look through the eyeglasses at arm's length and move them left, right, up, and down.
2. If objects seen through them move opposite to the motion of the glasses, the wearer is *farsighted.* The lenses are convex.
3. If objects seen through the eyeglasses move in the same direction as the glasses, then the wearer is *nearsighted.* The lenses are concave.

Of course, you must realize that there are many other reasons for wearing glasses. An optometrist uses many different lenses which are not discussed here. By the way, it is usually very difficult to tell whether eyeglass lenses are thicker or thinner at the center than at the edges just by feeling them.

SOUND AND LIGHT—
More to find out:

Will a loud noise blow out a candle?

You can demonstrate that air carries sound. Use a one-pound coffee can which has both ends removed. Stretch some rubber from a large balloon over one smooth end. Wrap the edges securely with string.

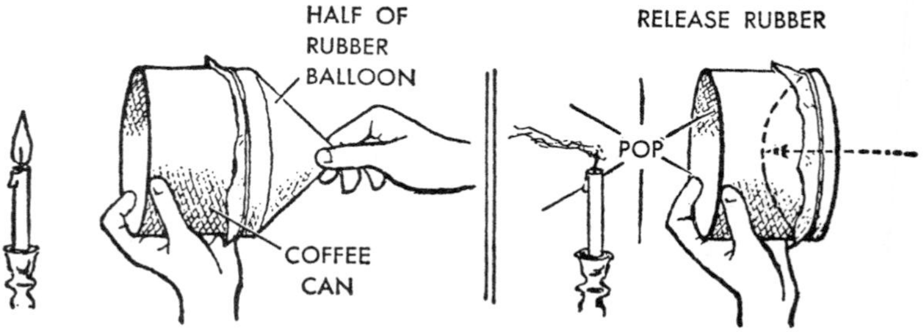

Face the open end of this drum toward a small lighted birthday candle. Pinch the center of the rubber membrane and pull it out. Release it and a loud boom will be heard. If properly placed, the candle flame will be blown out.

Can you blow over a card which is lightly held in an upright position? Does this explain why loud explosions often break nearby windows?

Do you want to hear chimes?

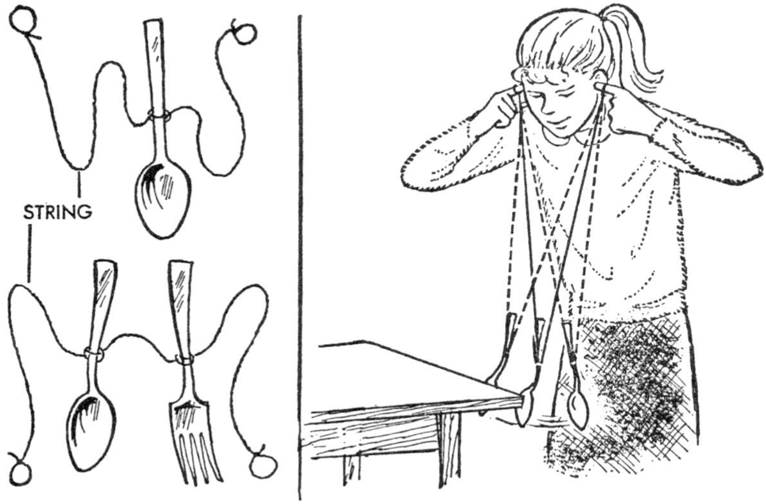

Get a piece of string about 3 feet long and tie small loops at each end. Tie a large spoon at the middle by means of a slip knot. Put a pointing finger from each hand a short distance into a loop and bring the fingers to each ear. Lean forward and swing the spoon so that it strikes the table. You will hear a deep pleasant sound resembling chimes. Touch the spoon and the vibration stops.

Experiment by listening to different sounds made when

spoons, forks, cups, metal pipes, and other metal objects are tied on separately. What happens when two objects are attached at one time and touch each other? What happens when space is left between them so that they cannot touch each other?

Can you "pipe" light?

Shine a flashlight beam into one end of a length of transparent plastic. It may illuminate objects held at the other end—even if the plastic is bent like a pretzel! Perhaps you have seen your dentist or doctor use such an instrument. The light is reflected over and over again along the length of the rod. Only a little light is lost along the way.

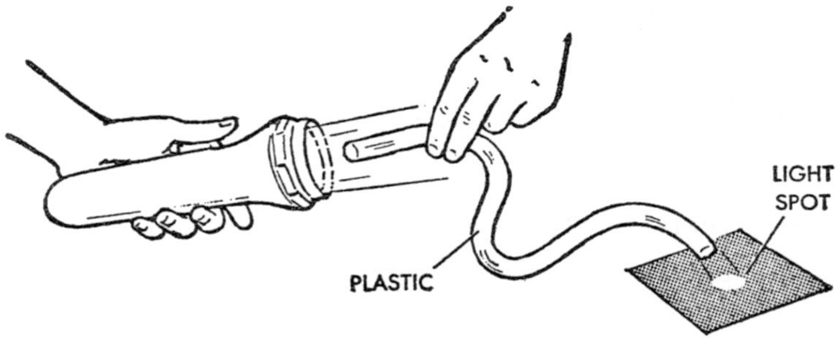

Test different types of solid plastic rods found in toothbrushes, towel and curtain rods, umbrella handles, ladies' pocketbook handles. Lucite and Plexiglass are trade names for the acrylic plastic best for this experiment.

A plastic rod can be bent into a curved shape by heating it in a baking oven until it softens. Use asbestos pot holders or other means to protect the fingers.

How can you see very tiny objects?

Sometimes you may wish to see a very small object clearly but you have no magnifying lens available. Or, if you must wear glasses to read, you may not have them with you when you need them. But if you have a pin and a piece of paper you can quickly make your own handy magnifier.

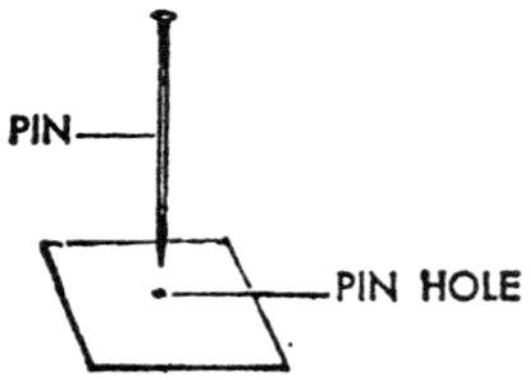

HOW TO READ FINE PRINT

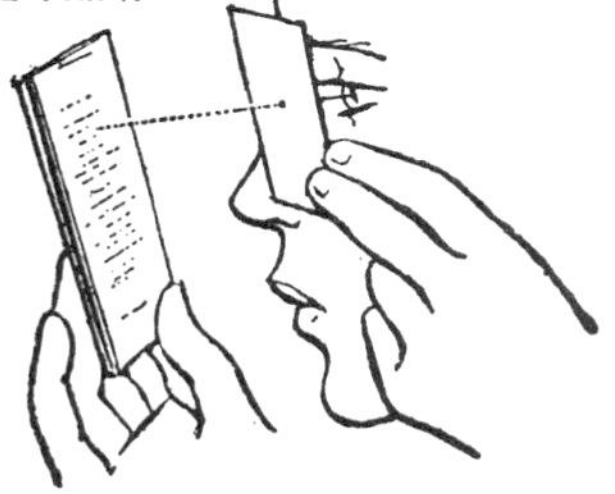

Make a small hole in the paper with the pin. Hold the paper close to what you wish to magnify or clarify and look through it with one eye. Notice the improvement. Since the tiny hole cuts down the light you may need stronger light for viewing.

Make viewers with many different hole sizes. What is the effect of each? Do you get better results with a clean round hole in aluminum foil? Also try cardboard, wood, and many other materials. Is it necessary to get a sharp focus with this magnifier?

Does shining a light in one eye affect the other?

When a strong light shines in your eye the iris (colored ring) closes down, making the pupil (opening) contract. This protects the eye. In dim light the pupil expands. If you look into a mirror and shine a flashlight into your eye you can see this movement.

Both eyes have adjusted themselves since birth so that they work intimately together. Suppose you shield one eye from the other, and shine the light only into one eye? Do you think the pupil of the unlighted eye will also contract? If so, how much?

In order to shield one eye from the other, place your outstretched palm tightly and vertically over your nose and forehead. Hold the flashlight close to the eye. Observe the effect on the lighted eye. Then look at the shaded eye, while shining the light on the other eye.

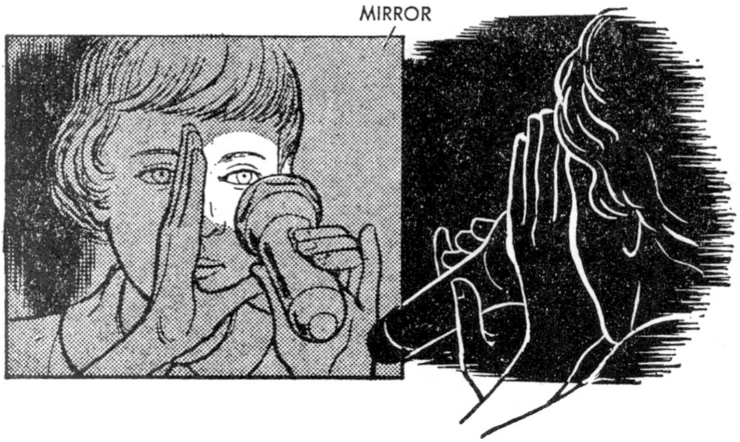

MEASUREMENT

Can you measure one thousandth of an inch?

By means of very simple mathematics and a small screw clamp, or even a nut and a bolt, you can measure things with amazing precision. You can be accurate to 1/100 of an inch every time. With only a little more care you can get fair results to within several thousandths of an inch.

A small screw clamp which operates smoothly is all that is needed. Study it and you will find that when the threaded bolt is made to rotate one complete turn, it moves into the clamp a distance equal to the space between the threads. This distance between threads is called the PITCH. (See illus., p. 142.)

The first thing you must do is to calculate the pitch accurately. One way is to place a good ruler alongside the threads and count the number of threads in 1 inch. Running a sharp pencil or pin along the bolt and listening to the clicks will help your accuracy.

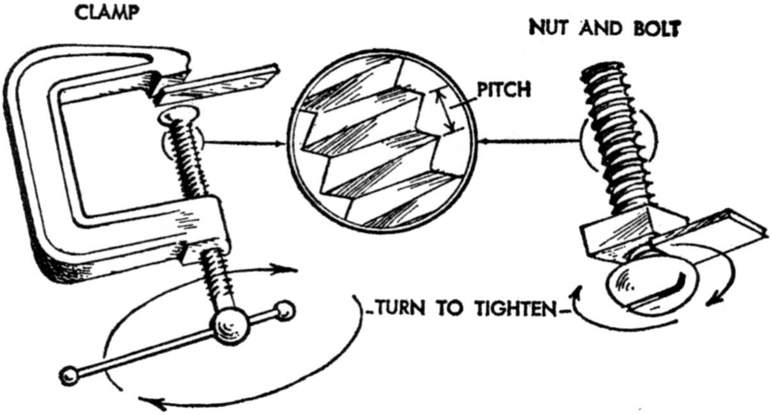

There is another way to find out the number of threads in 1 inch. Tighten the clamp on an object which is *definitely known to be 1 inch.* Remove the object and count the number of complete turns of the handle needed to close the clamp. You can use both methods as a check on yourself. Do each several times until you feel that you have the true number of threads per inch.

To calculate the pitch, simply divide the number of threads into 1 inch. For example, if you counted 24 threads to 1 inch, then the pitch is $\frac{1}{24}$ of an inch. In other words, each complete turn of the handle advances the bolt $\frac{1}{24}$ of an inch.

You are now ready to measure something small, such as a thin sheet of metal. First tighten it in the clamp with the least amount of force necessary to hold it in place. Observe carefully the angle of the handle. Then, loosen slightly in order to release the object being measured. Now bring the handle back to where it was.

Using this as a starting place, count the number of turns needed to close the clamp. Suppose it takes one complete turn. Then the metal is $\frac{1}{24}$ of an inch thick. If it takes two complete turns to close the clamp, then the metal is 2 x $\frac{1}{24}$ inch thick. However, if it takes two and one-half turns, then your calculations are: $2\frac{1}{2}$ x $\frac{1}{24}$.

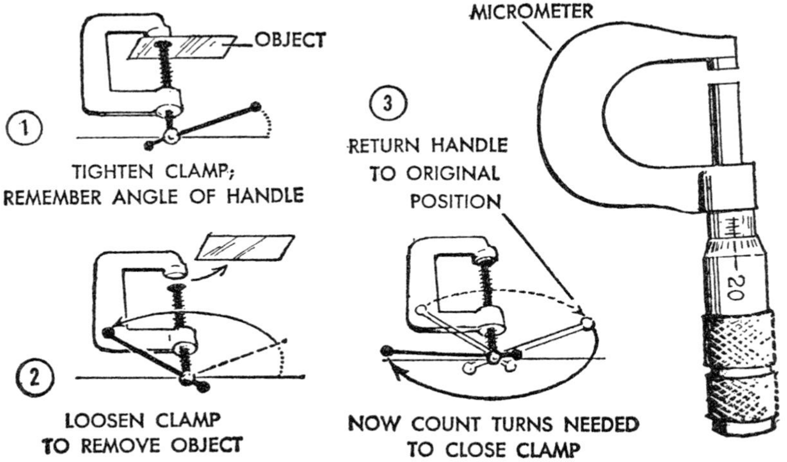

OBJECT
1
TIGHTEN CLAMP;
REMEMBER ANGLE OF HANDLE
2
LOOSEN CLAMP
TO REMOVE OBJECT
3
RETURN HANDLE
TO ORIGINAL
POSITION
NOW COUNT TURNS NEEDED
TO CLOSE CLAMP
MICROMETER
20

Work out the multiplication of fractions problem to the third decimal place. This will give you the thickness in thousands of an inch.

For example:

$$2\tfrac{1}{2} \times \tfrac{1}{24} = \tfrac{5}{2} \times \tfrac{1}{24} = \tfrac{5}{48} = .104 \text{ inches}$$

The result is also referred to as 104 thousands of an inch.

An instrument very similar to this is used by scientists and machinists for measuring very small distances with great accuracy. It is called a MICROMETER (my-KROM-it-er).

If you wish to get more precise results with your home-made micrometer, get a clamp with finer threads. Another hint is to use a protractor for getting fractional turns of the handle. Suppose it takes four complete turns and then 55 degrees beyond the starting point to close the clamp.

Since 360 degrees represent one complete turn then 55 degrees is $\frac{55}{360}$ of one turn. Therefore, the total number of turns is $4\frac{55}{360}$. Multiply this by the pitch to obtain your measurement. Of course, no one expects you to be accurate to within 1 degree. However, using a protractor is better than guessing the fraction of the turn.

You can check your instrument by seeing how close your measurement comes to the known thickness of the following objects: No. 18 wire (ordinary bell wire) is .040 inch; No. 22 wire is .025 inch; ⅛-inch metal drill is .125 inch; ¼-inch metal drill is .250 inch.

Perhaps someone you know has a micrometer. Check your instrument against his. See how many thousandths of an inch you differ from his measurements. Both of you will be pleasantly surprised at the degree of accuracy.

Are gas range pilot lights expensive?

Most people find it a great convenience not to have to strike matches to light a gas range burner. But it would annoy them if they had to pay much more than a dollar a month for this luxury.

Some modern gas stoves may have half a dozen pilot lights which consume gas continuously. When the gas bill arrives, the housewife may wonder why the monthly charge is so great. How much does it cost your family to maintain such pilot lights? Your mother should appreciate the answer to your research problem.

In practically all homes where gas is piped in, there is a gas meter containing many dials. Each dial clearly indicates how many cubic feet of gas will cause the pointer to make one revolution.

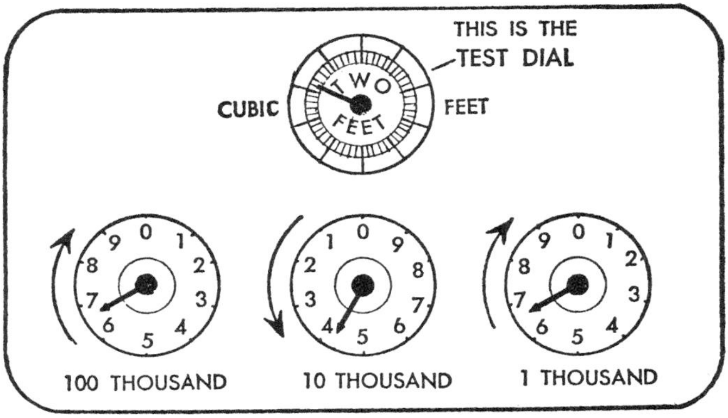

Your meter also has one test dial. This is usually marked like one of the following: HALF FOOT PER REV., TWO FEET PER REV., or FIVE FEET PER REV. For your tests then, use the dial on your meter with the lowest volume, since its pointer will move faster than the others.

Choose a time of day when your mother is not cooking. Make certain that only the pilot lights are on in the stove. If the meter is in a dark area use a flashlight or nearby electric light to illuminate the meter and your watch. Wait until the pointer is on a line on the test dial. Jot down on your note paper the exact time and also, a diagram of your starting mark as seen on the dial.

Record the time it takes for the pointer to make one revolution, that is, to come back to the starting line. It is a good experimental procedure to do this several times, until you are convinced that your timing of one revolution is accurate. You may wish to average the results of several revolutions.

You are now ready for the arithmetic of this problem. It is very simple when taken step by step:

a) There are 43,200 minutes in one month; (60 minutes in one hour × 24 hours in a day × 30 days in a month = 43,200)

b) Divide 43,200 by the number of minutes it takes for the test pointer to make one revolution. This gives you *the number of revolutions of the test pointer in one month* caused by the use of the pilot lights.

c) Multiply this number of monthly revolutions of the test pointer by the type of test dial you used. This gives you *the number of cubic feet of gas used in one month by your pilot lights.*

d) Multiply this number by the cost of *one* cubic foot of gas. This is usually shown in some way on your gas bill. If not, ask your gas company for the rates.

Let us do a problem. Suppose it takes 144 minutes for the pointer of a 2-cubic-foot test dial to make one revolution, when only pilot lights are burning. The gas rate in your town is \$1.20 per 1,000 cubic feet (HINT: The cost of 1 cubic foot is $\frac{1.20}{1,000}$, which equals \$.0012.) What is the monthly cost of the pilots?

$$\frac{43,200}{144} \times 2 \times \$.0012 = \$.72$$

The cost of the pilot lights is \$.72 per month.

MEASUREMENT—*More to find out:*

How can the height of a big hill be measured?

Borrow an aneroid barometer. You may have seen this "weather forecaster" hanging on a wall. On the dial is printed RAIN—CHANGE—FAIR or similar words. This instrument indicates changes in air pressure. It can also be used to measure height above the ground, because the air pressure goes down 1 inch for every 900 feet rise in elevation.

Go to the base of a hill or mountain and write down the pressure as indicated by the bottom of the two pointers. You may also set the top pointer over the bottom one. It does not move and only serves as a reminder by indicating the last reading. Carry the instrument carefully and without jarring to the top of the hill. Read the pressure again. Subtract from the former pressure. (Watch the decimal point!)

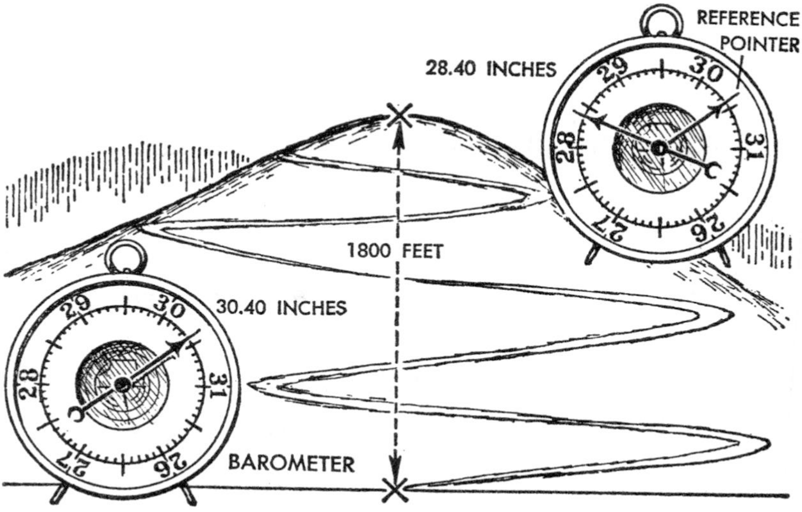

When you multiply the air pressure difference expressed in inches by 900, you will know the number of feet of elevation. Check your answer by looking at a local topographical map. Try this experiment while going up or down in an elevator of a tall building. It should also show a difference in pressures between the basement of your school and the top floor.

Needless to say, this method of finding altitude may not be accurate if an interval of many hours elapses between barometer readings. This is because the air pressure may have changed during this time because of weather conditions.

Does a rubber band stretch evenly?

If a rubber band stretches 1 inch when it is pulled by a certain force, will it stretch 2 inches when the force is doubled? Will it stretch 20 inches when the force is increased twenty times? Will a rubber band two times as thick as another hold up two times as much weight before it breaks? You can answer these, and many other questions by means of a simple experimental device.

Get a 6-inch length of thin rubber by breaking a suitable rubber band. Look around for some place where the rubber can be suspended while it is being stretched by weights. For example, one end can be tied to a thumb tack which is stuck into the edge of an overhanging shelf or ledge.

Tie the lower end to the head of a very large nail. Tape all knots, since they have a tendency to unravel when pulled. With a yardstick measure the distance the nail head is from the top of the length of rubber. Record this in your notes.

Now attach a *similar* nail to the first suspended nail by means of a short strip of cellophane or masking tape. Again record the length. Keep adding single nails of the same kind, and measuring the new lengths until the elastic breaks. Repeat the experiment with similar rubber bands. Are the results almost the same?

Make charts and graphs of your results. Think of many other experiments to try. You may substitute other objects for the nails, as long as all objects have the same weight.

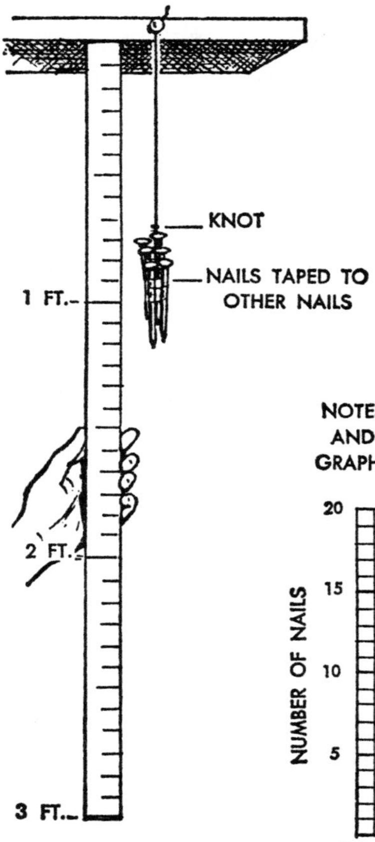

NUMBER OF NAILS	STRETCH IN INCHES
1	5
2	5½
3	6
4	6½
5	7¼
6	8¼
7	9¼
8	10½
9	11½
10	12¾
11	14
12	15
13	16
14	17
15	17¾
16	18¾
17	19½
18	20⅛
19	20½
20	20⅞

NOTES AND GRAPH

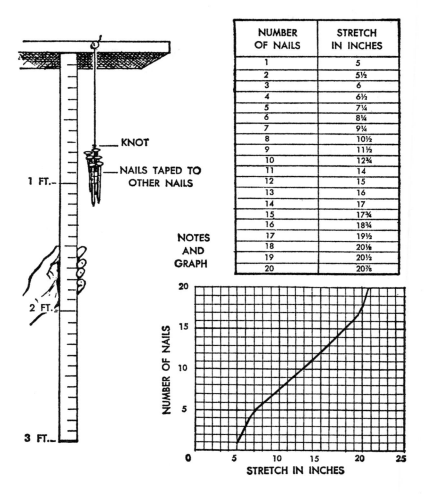

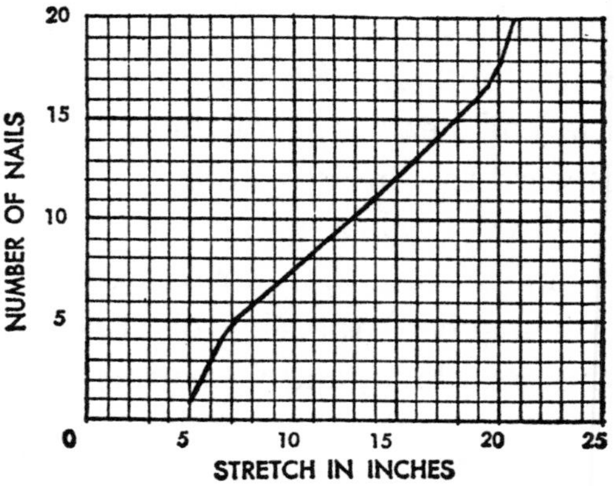

Can you make an accurate postal scale?

By building a simple scale you can weigh your heavy letters and paste the correct postage on them. With a little thought you can improve it to weigh other light objects.

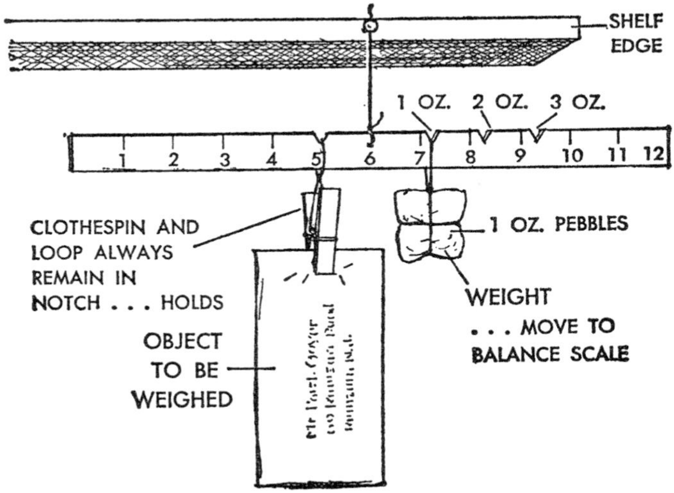

You will need a 12-inch ruler or a similar strip of wood. Drill a fine pinhole at the 6-inch mark quite close to one edge. A hole can also be made by forcing through the wood a red-hot pin held by pliers. Tie the end of a sewing thread through this hole. Devise some way to suspend the ruler a short distance from a shelf or table edge.

Get four 1-ounce weights by weighing small pebbles in envelopes at your grocer, butcher, or pharmacist.

Fold each envelope into a small package and tie it. Attach a loop of thread to a 1-ounce envelope and another loop of thread to a spring clothes pin. The loops should be able to slip over the ruler.

Clasp a 1-ounce weight in the clothes pin. Slip the loop of the clothes pin into a small notch or slit made over the 5-inch mark on the ruler. Slip the loop of the other 1-ounce weight over the ruler at about the 7-inch mark. Move the loop left or right until the ruler is horizontal. Use the shelf as a reference line. Make a small notch in the ruler to hold the loop at the correct spot. Print 1 oz. under this notch. Remove the 1-ounce weight from the clothes pin. After this, the letter to be weighed will be held by the clothes pin. You will easily see whether it is underweight or overweight.

If a known 2-ounce weight is now placed in the clothes pin, the 1-ounce balancing weight must be moved to the right until the ruler is horizontal again. Make a small notch at the correct place and print 2 ozs. under it.

Continue placing 3, 4, and 5 ounces in the clothes pin, adjusting the loop of the balancing weight to the right, notching and marking the ruler. The clothes pin loop, of course, must always be in the same place.

You can discover many other uses of this device besides weighing letters. For example, you can have a pan suspended from the loop instead of a clothes pin. Then you can weigh powders. If you wish to weigh fractions of ounces, mark off equal distances between ounces.

How far from a wall must you be to hear an echo?

Face a wall known to produce good echoes. Stand about 5 feet away and make a short loud sound. Do you hear an echo? Keep repeating the sound while backing away. Stop when you hear the echo and measure your distance from the wall. Is it about 56 feet?

Your ears can only distinguish two sounds as separate when they are at least one-tenth of a second apart. Sound travels about 1,120 feet per second, or 112 feet in one-tenth of a second. Therefore, to produce an echo, sound has to travel at least 112 feet. It covers that round-trip distance when you are 56 feet away from the wall.

Here Are Some Good Science Books for More Ideas

Marion Baer, *Sound: An Experiment Book*, New York: Holiday House, Inc., 1952.

George Barr, *Outdoor Science Projects for Young People*, New York: Dover Publications, Inc., 1991.

———, *Science Research Experiments for Young People*, New York: Dover Publications, Inc., 1989.

———, *Young Scientist Takes a Ride*, New York: McGraw-Hill, Inc., 1960.

Nelson Beeler and Franklyn Branley, *Experiments in Chemistry*, New York: Thomas Y. Crowell Company, 1952.

———, *Experiments in Science*, New York: Thomas Y. Crowell Company, 1955.

———, *Experiments with Electricity*, New York: Thomas Y. Crowell Company, 1949.

Mae and Ira Freeman, *Fun with Chemistry*, New York: Random House, 1944.

———, *Fun with Astronomy*, New York: Random House, 1953.

Harold Gatty, *Nature Is Your Guide: How to Find Your Way on Land and Sea by Observing Nature*, New York: E. P. Dutton and Company, Inc., 1957.

C. J. Lynde, *Science Experiences with Home Equipment*, 3rd Ed., Princeton, N.J.: D. Van Nostrand Company, Inc., 1949.

———, *Science Experiences with Inexpensive Equipment*, 3rd Ed., Princeton, N.J.: D. Van Nostrand Company, Inc., 1950.

Alfred Morgan, *Things a Boy Can Do with Electricity*, New York: Charles Scribner's Sons, 1953.

Hy Ruchlis, *Orbit*, New York: Harper and Brothers, 1958.

———, *The Wonder of Light*, New York: Harper and Brothers, 1960.

Herman Schneider, *Everyday Machines and How They Work*, New York: McGraw-Hill, Inc., 1950.
———, *Everyday Weather and How It Works*, Rev. Ed. New York: McGraw-Hill, Inc., 1961.
Herman and Nina Schneider, *How Your Body Works*, New York: William R. Scott, Inc., 1949.
Leo Schneider, *You and Your Senses*, New York: Harcourt, Brace and Company, Inc., 1965.
Julius Schwartz, *It's Fun to Know Why*, New York: McGraw-Hill, Inc., 1952.
———, *Through the Magnifying Glass*, New York: McGraw-Hill, Inc., 1954.
K. M. Swezey, *After-dinner Science*, revised edition, New York: McGraw-Hill, Inc., 1961.
———, *Science Magic*, New York: McGraw-Hill, Inc., 1952.
Edwin Way Teale, *The Junior Book of Insects*, Rev. Ed., New York: E. P. Dutton and Company, 1953.
Rose Wyler and Gerald Ames, *The Golden Book of Astronomy*, Rev. Ed., New York: Golden Press, 1959.

Index

A CATALOG OF SELECTED

DOVER BOOKS

IN ALL FIELDS OF INTEREST

ILLUSTRATED GUIDE TO SHAKER FURNITURE, Robert Meader. All furniture and appurtenances, with much on unknown local styles. 235 photos. 146pp. 9 × 12. 22819-3 Pa. $8.95

WHALE SHIPS AND WHALING: A Pictorial Survey, George Francis Dow. Over 200 vintage engravings, drawings, photographs of barks, brigs, cutters, other vessels. Also harpoons, lances, whaling guns, many other artifacts. Comprehensive text by foremost authority. 207 black-and-white illustrations. 288pp. 6 × 9. 24808-9 Pa. $9.95

THE BERTRAMS, Anthony Trollope. Powerful portrayal of blind self-will and thwarted ambition includes one of Trollope's most heartrending love stories. 497pp. 5⅜ × 8½. 25119-5 Pa. $9.95

ADVENTURES WITH A HAND LENS, Richard Headstrom. Clearly written guide to observing and studying flowers and grasses, fish scales, moth and insect wings, egg cases, buds, feathers, seeds, leaf scars, moss, molds, ferns, common crystals, etc.—all with an ordinary, inexpensive magnifying glass. 209 exact line drawings aid in your discoveries. 220pp. 5⅜ × 8½. 23330-8 Pa. $4.95

RODIN ON ART AND ARTISTS, Auguste Rodin. Great sculptor's candid, wide-ranging comments on meaning of art; great artists; relation of sculpture to poetry, painting, music; philosophy of life, more. 76 superb black-and-white illustrations of Rodin's sculpture, drawings and prints. 119pp. 8⅜ × 11¼. 24487-3 Pa. $7.95

FIFTY CLASSIC FRENCH FILMS, 1912–1982: A Pictorial Record, Anthony Slide. Memorable stills from Grand Illusion, Beauty and the Beast, Hiroshima, Mon Amour, many more. Credits, plot synopses, reviews, etc. 160pp. 8¼ × 11. 25256-6 Pa. $11.95

THE PRINCIPLES OF PSYCHOLOGY, William James. Famous long course complete, unabridged. Stream of thought, time perception, memory, experimental methods; great work decades ahead of its time. 94 figures. 1,391pp. 5⅜ × 8½. 20381-6, 20382-4 Pa., Two-vol. set $23.90

BODIES IN A BOOKSHOP, R. T. Campbell. Challenging mystery of blackmail and murder with ingenious plot and superbly drawn characters. In the best tradition of British suspense fiction. 192pp. 5⅜ × 8½. 24720-1 Pa. $4.95

CALLAS: PORTRAIT OF A PRIMA DONNA, George Jellinek. Renowned commentator on the musical scene chronicles incredible career and life of the most controversial, fascinating, influential operatic personality of our time. 64 black-and-white photographs. 416pp. 5⅜ × 8¼. 25047-4 Pa. $8.95

GEOMETRY, RELATIVITY AND THE FOURTH DIMENSION, Rudolph Rucker. Exposition of fourth dimension, concepts of relativity as Flatland characters continue adventures. Popular, easily followed yet accurate, profound. 141 illustrations. 133pp. 5⅜ × 8½. 23400-2 Pa. $4.95

HOUSEHOLD STORIES BY THE BROTHERS GRIMM, with pictures by Walter Crane. 53 classic stories—Rumpelstiltskin, Rapunzel, Hansel and Gretel, the Fisherman and his Wife, Snow White, Tom Thumb, Sleeping Beauty, Cinderella, and so much more—lavishly illustrated with original 19th century drawings. 114 illustrations. x + 269pp. 5⅜ × 8½. 21080-4 Pa. $4.95

PLANTS OF THE BIBLE, Harold N. Moldenke and Alma L. Moldenke. Standard reference to all 230 plants mentioned in Scriptures. Latin name, biblical reference, uses, modern identity, much more. Unsurpassed encyclopedic resource for scholars, botanists, nature lovers, students of Bible. Bibliography. Indexes. 123 black-and-white illustrations. 384pp. 6 × 9. 25069-5 Pa. $8.95

FAMOUS AMERICAN WOMEN: A Biographical Dictionary from Colonial Times to the Present, Robert McHenry, ed. From Pocahontas to Rosa Parks, 1,035 distinguished American women documented in separate biographical entries. Accurate, up-to-date data, numerous categories, spans 400 years. Indices. 493pp. 6½ × 9¼. 24523-3 Pa. $10.95

THE FABULOUS INTERIORS OF THE GREAT OCEAN LINERS IN HISTORIC PHOTOGRAPHS, William H. Miller, Jr. Some 200 superb photographs capture exquisite interiors of world's great "floating palaces"—1890s to 1980s: *Titanic, Ile de France, Queen Elizabeth, United States, Europa,* more. Approx. 200 black-and-white photographs. Captions. Text. Introduction. 160pp. 8⅞ × 11¾. 24756-2 Pa. $9.95

THE GREAT LUXURY LINERS, 1927–1954: A Photographic Record, William H. Miller, Jr. Nostalgic tribute to heyday of ocean liners. 186 photos of *Ile de France, Normandie, Leviathan, Queen Elizabeth, United States,* many others. Interior and exterior views. Introduction. Captions. 160pp. 9 × 12. 24056-8 Pa. $12.95

A NATURAL HISTORY OF THE DUCKS, John Charles Phillips. Great landmark of ornithology offers complete detailed coverage of nearly 200 species and subspecies of ducks: gadwall, sheldrake, merganser, pintail, many more. 74 full-color plates, 102 black-and-white. Bibliography. Total of 1,920pp. 8⅝ × 11¼. 25141-1, 25142-X Cloth., Two-vol. set $100.00

THE SEAWEED HANDBOOK: An Illustrated Guide to Seaweeds from North Carolina to Canada, Thomas F. Lee. Concise reference covers 78 species. Scientific and common names, habitat, distribution, more. Finding keys for easy identification. 224pp. 5⅜ × 8½. 25215-9 Pa. $6.95

THE TEN BOOKS OF ARCHITECTURE: The 1755 Leoni Edition, Leon Battista Alberti. Rare classic helped introduce the glories of ancient architecture to the Renaissance. 68 black-and-white plates. 336pp. 8⅜ × 11¼. 25239-6 Pa. $14.95

MISS MACKENZIE, Anthony Trollope. Minor masterpieces by Victorian master unmasks many truths about life in 19th-century England. First inexpensive edition in years. 392pp. 5⅜ × 8½. 25201-9 Pa. $8.95

THE RIME OF THE ANCIENT MARINER, Gustave Doré, Samuel Taylor Coleridge. Dramatic engravings considered by many to be his greatest work. The terrifying space of the open sea, the storms and whirlpools of an unknown ocean, the ice of Antarctica, more—all rendered in a powerful, chilling manner. Full text. 38 plates. 77pp. 9¼ × 12. 22305-1 Pa. $4.95

THE EXPEDITIONS OF ZEBULON MONTGOMERY PIKE, Zebulon Montgomery Pike. Fascinating firsthand accounts (1805–6) of exploration of Mississippi River, Indian wars, capture by Spanish dragoons, much more. 1,088pp. 5⅜ × 8½. 25254-X, 25255-8 Pa., Two-vol. set $25.90

DEGAS: An Intimate Portrait, Ambroise Vollard. Charming, anecdotal memoir by famous art dealer of one of the greatest 19th-century French painters. 14 black-and-white illustrations. Introduction by Harold L. Van Doren. 96pp. 5⅜ × 8½.
25131-4 Pa. $4.95

PERSONAL NARRATIVE OF A PILGRIMAGE TO AL-MADINAH AND MECCAH, Richard F. Burton. Great travel classic by remarkably colorful personality. Burton, disguised as a Moroccan, visited sacred shrines of Islam, narrowly escaping death. 47 illustrations. 959pp. 5⅜ × 8½.
21217-3, 21218-1 Pa., Two-vol. set $19.90

PHRASE AND WORD ORIGINS, A. H. Holt. Entertaining, reliable, modern study of more than 1,200 colorful words, phrases, origins and histories. Much unexpected information. 254pp. 5⅜ × 8½. 20758-7 Pa. $5.95

THE RED THUMB MARK, R. Austin Freeman. In this first Dr. Thorndyke case, the great scientific detective draws fascinating conclusions from the nature of a single fingerprint. Exciting story, authentic science. 320pp. 5⅜ × 8½. (Available in U.S. only) 25210-8 Pa. $6.95

AN EGYPTIAN HIEROGLYPHIC DICTIONARY, E. A. Wallis Budge. Monumental work containing about 25,000 words or terms that occur in texts ranging from 3000 B.C. to 600 A.D. Each entry consists of a transliteration of the word, the word in hieroglyphs, and the meaning in English. 1,314pp. 6⅞ × 10.
23615-3, 23616-1 Pa., Two-vol. set $35.90

THE COMPLEAT STRATEGYST: Being a Primer on the Theory of Games of Strategy, J. D. Williams. Highly entertaining classic describes, with many illustrated examples, how to select best strategies in conflict situations. Prefaces. Appendices. xvi + 268pp. 5⅜ × 8½. 25101-2 Pa. $6.95

THE ROAD TO OZ, L. Frank Baum. Dorothy meets the Shaggy Man, little Button-Bright and the Rainbow's beautiful daughter in this delightful trip to the magical Land of Oz. 272pp. 5⅜ × 8. 25208-6 Pa. $5.95

POINT AND LINE TO PLANE, Wassily Kandinsky. Seminal exposition of role of point, line, other elements in nonobjective painting. Essential to understanding 20th-century art. 127 illustrations. 192pp. 6½ × 9¼. 23808-3 Pa. $5.95

LADY ANNA, Anthony Trollope. Moving chronicle of Countess Lovel's bitter struggle to win for herself and daughter Anna their rightful rank and fortune—perhaps at cost of sanity itself. 384pp. 5⅜ × 8½. 24669-8 Pa. $8.95

EGYPTIAN MAGIC, E. A. Wallis Budge. Sums up all that is known about magic in Ancient Egypt: the role of magic in controlling the gods, powerful amulets that warded off evil spirits, scarabs of immortality, use of wax images, formulas and spells, the secret name, much more. 253pp. 5⅜ × 8½. 22681-6 Pa. $4.95

THE DANCE OF SIVA, Ananda Coomaraswamy. Preeminent authority unfolds the vast metaphysic of India: the revelation of her art, conception of the universe, social organization, etc. 27 reproductions of art masterpieces. 192pp. 5⅜ × 8½.
24817-8 Pa. $6.95

CHRISTMAS CUSTOMS AND TRADITIONS, Clement A. Miles. Origin, evolution, significance of religious, secular practices. Caroling, gifts, yule logs, much more. Full, scholarly yet fascinating; non-sectarian. 400pp. 5⅜ × 8½. 23354-5 Pa. $7.95

THE HUMAN FIGURE IN MOTION, Eadweard Muybridge. More than 4,500 stopped-action photos, in action series, showing undraped men, women, children jumping, lying down, throwing, sitting, wrestling, carrying, etc. 390pp. 7⅞ × 10⅝. 20204-6 Cloth. $24.95

THE MAN WHO WAS THURSDAY, Gilbert Keith Chesterton. Witty, fast-paced novel about a club of anarchists in turn-of-the-century London. Brilliant social, religious, philosophical speculations. 128pp. 5⅜ × 8½. 25121-7 Pa. $3.95

A CÉZANNE SKETCHBOOK: Figures, Portraits, Landscapes and Still Lifes, Paul Cézanne. Great artist experiments with tonal effects, light, mass, other qualities in over 100 drawings. A revealing view of developing master painter, precursor of Cubism. 102 black-and-white illustrations. 144pp. 8¾ × 6⅜. 24790-2 Pa. $6.95

AN ENCYCLOPEDIA OF BATTLES: Accounts of Over 1,560 Battles from 1479 B.C. to the Present, David Eggenberger. Presents essential details of every major battle in recorded history, from the first battle of Megiddo in 1479 B.C. to Grenada in 1984. List of Battle Maps. New Appendix covering the years 1967–1984. Index. 99 illustrations. 544pp. 6½ × 9¼. 24913-1 Pa. $14.95

AN ETYMOLOGICAL DICTIONARY OF MODERN ENGLISH, Ernest Weekley. Richest, fullest work, by foremost British lexicographer. Detailed word histories. Inexhaustible. Total of 856pp. 6½ × 9¼. 21873-2, 21874-0 Pa., Two-vol. set $19.90

WEBSTER'S AMERICAN MILITARY BIOGRAPHIES, edited by Robert McHenry. Over 1,000 figures who shaped 3 centuries of American military history. Detailed biographies of Nathan Hale, Douglas MacArthur, Mary Hallaren, others. Chronologies of engagements, more. Introduction. Addenda. 1,033 entries in alphabetical order. xi + 548pp. 6½ × 9¼. (Available in U.S. only) 24758-9 Pa. $13.95

LIFE IN ANCIENT EGYPT, Adolf Erman. Detailed older account, with much not in more recent books: domestic life, religion, magic, medicine, commerce, and whatever else needed for complete picture. Many illustrations. 597pp. 5⅜ × 8½. 22632-8 Pa. $9.95

HISTORIC COSTUME IN PICTURES, Braun & Schneider. Over 1,450 costumed figures shown, covering a wide variety of peoples: kings, emperors, nobles, priests, servants, soldiers, scholars, townsfolk, peasants, merchants, courtiers, cavaliers, and more. 256pp. 8⅜ × 11¼. 23150-X Pa. $9.95

THE NOTEBOOKS OF LEONARDO DA VINCI, edited by J. P. Richter. Extracts from manuscripts reveal great genius; on painting, sculpture, anatomy, sciences, geography, etc. Both Italian and English. 186 ms. pages reproduced, plus 500 additional drawings, including studies for *Last Supper, Sforza* monument, etc. 860pp. 7⅞ × 10¾. (Available in U.S. only) 22572-0, 22573-9 Pa., Two-vol. set $35.90